Intracranial Epidural Bleeding

Intracranial Epidural Bleeding

History, Management, and Pathophysiology

Jeremy Christopher Ganz
Haukeland University Hospital, Bergen, Norway

ACADEMIC PRESS

An imprint of Elsevier

Academic Press is an imprint of Elsevier
125 London Wall, London EC2Y 5AS, United Kingdom
525 B Street, Suite 1800, San Diego, CA 92101-4495, United States
50 Hampshire Street, 5th Floor, Cambridge, MA 02139, United States
The Boulevard, Langford Lane, Kidlington, Oxford OX5 1GB, United Kingdom

Notices
Knowledge and best practice in this field are constantly changing. As new research and experience broaden our
understanding, changes in research methods, professional practices, or medical treatment may become
necessary.

Practitioners and researchers must always rely on their own experience and knowledge in evaluating and using
any information, methods, compounds, or experiments described herein. In using such information or methods
they should be mindful of their own safety and the safety of others, including parties for whom they have a
professional responsibility.

To the fullest extent of the law, neither the Publisher nor the authors, contributors, or editors, assume any
liability for any injury and/or damage to persons or property as a matter of products liability, negligence or
otherwise, or from any use or operation of any methods, products, instructions, or ideas contained in the
material herein.

British Library Cataloguing-in-Publication Data
A catalogue record for this book is available from the British Library

Library of Congress Cataloging-in-Publication Data
A catalog record for this book is available from the Library of Congress

ISBN: 978-0-12-812159-7

For Information on all Academic Press publications
visit our website at https://www.elsevier.com/books-and-journals

  **Working together
to grow libraries in
developing countries**

www.elsevier.com • www.bookaid.org

Publisher: Nikki Levy
Acquisition Editor: Melanie Tucker
Editorial Project Manager: Kristi Anderson
Production Project Manager: Mohana Natarajan
Cover Designer: Alan Studholme

Typeset by MPS Limited, Chennai, India

DEDICATION

Prof. Lindsay Symon was one of my first clinical chiefs in neurosurgery and my first research director. Lindsay has been a towering figure in both British and international neurosurgical circles, becoming the President of the International Association of Neurosurgical Societies. I remember him for his kindness and humor. In my time at Queen Square I heard him lecture and he probably has no equal as an academic lecturer. On moving to Norway, it was natural for Scandinavian colleagues to approach him for an assessment and it is with deep gratitude that I can note that he was invariably loyal and supportive. Thus, it is with great pleasure that I dedicate this book to him.

CONTENTS

Epidural hematomas are a source of frustration. In principle, they should be easy to treat but in practice even today, this is not always the case. They have been recognized possibly as early as 2500 years ago by Hippocrates himself. Moreover, it would seem that in the 1st-century AD Celsus was aware that such a hematoma could form in the absence of a cranial fracture.

Over the succeeding two millennia, it is possible to trace many concepts, some correct and some erroneous which form the twisting pathway of knowledge. The first part of this book relates the steps in this pathway backwards and forwards from antiquity to the Renaissance and subsequently to the late 19th century when modern ideas finally emerged.

The second section covers the evolving clinical management and increasingly effective methods of investigation. Investigations are now so effective that they could be characterized as close to optimal. With this in mind, the current difficulties in management are analyzed with a view to suggesting ways of still further improvement.

Finally, attention is directed to the rather complex pathophysiology which underlies the formation of these entities. There is rather more known about this aspect of epidural bleeding than is often appreciated and the book should hopefully serve to make the relevant principles more widely understood.

ACKNOWLEDGMENTS

The debts involved in the writing of this book go back to the beginning of the author's career when he was in training. Valentine Logue and Lindsay Symon started him on the road to respectively clinical assessment and scientific analysis. He received thorough training in operative technique from Douglas Phillips, Allan Hulme, Huw Griffith, and Brian Cummins in Bristol. In Manchester, he was privileged to be exposed to the matchless operative technique of Richard Johnson. These teachers provided background from which the relevance of scientific work may be rightly judged. In relation to this book, the author owes a debt of gratitude before all others to the late Nic Zwetnow; his mentor during PhD studies. Nic's imagination, intellectual rigor, and generosity remain an unrepayable debt. The physiological studies about epidural studies benefited much from discussions with Prof. Zwetnow's other doctoral students, not least Jan Orlin and the much missed, late Slobodan Vlaikovic. The late Urban Ponten in Uppsala also gave wise advice especially regarding the MRI studies.

At Haukeland Hospital in Bergen thanks are due to Prof. Erik-Olof Backlund, not least for his insistence on high academic standards. He made sure time was available for me to continue my studies. Prof. Johan Aarli, the head of the Department of Neurology and subsequently the President of the World Federation of Neurology was a constant source of stimulation.

The earliest chapters are concerned with the classical world from Hippocrates to Galen. The author is grateful for discussions with and advice from Sir Geoffrey Lloyd, the Emeritus Professor of Ancient Philosophy and Science in the University of Cambridge. Further valuable advice was given by Prof. Boleslav Lichterman of the I.M. Sechenov First Moscow State Medical University of the Ministry of Healthcare of the Russian Federation. A particular thanks is due to Maheep Singh Gaur without whose assistance chapter 16 would not have been possible.

The author is also grateful to the numerous agencies which have made available digitized versions of original medical texts on the internet. This is an immense resource without which this book could not have been written.

Any author, no matter what kind of writing is involved will know the debt we owe to our partners, in my case my wife Annie Gao. She has tolerated all the changing situations, absences, and other vicissitudes occasioned by the writing of this book and without her support and help it would never have been completed.

SECTION *I*

Epidural Hematoma—
Relevant Basic Knowledge

CHAPTER *1*

Introduction

INTRODUCTION

To the best of the author's knowledge no definitive history of epidural hematomas and their management has been written to date. The first systematic modern account of epidural bleeding was published in 1886.[1] Yet there must have been a number of them on the battlefields of the ancient world but a clear description as opposed to text which might just refer to such bleeding is not available. From the battlefields of ancient Greece, it took over 2000 years before it could be stated that a conceptual framework existed which could form a basis for rational treatment.

The following elements are necessary for the development of an epidural hematoma. Fig. 1.1 shows the basic anatomical components. For an epidural hematoma to form the dura must be detached from the skull to form a space into which blood may accumulate. To modern eyes this all seems very obvious. Yet, following the earliest identification of the brain and its surrounding membranes it took millennia rather than centuries for these ideas to form.

The necessary components of understanding are of course normal and pathological anatomy, physiology, and their clinical expression. This can be split up into different components. All of this will seem very simple to medical professionals but it is necessary to list up these components if one is to gain an understanding of how modern knowledge came to develop. This is by no means a simple process. There is no need to consider all aspects of brain function only those that relate to epidural bleeding.

1. **The Skull**

   This is a hollow bony container made up of a number of bones which all are attached to each other rather like the pieces of a jigsaw but with much finer processes going in and out. The locations of the attachments are called sutures. The variable appearance of cranial sutures is shown in Fig. 1.2.

Intracranial Epidural Bleeding. DOI: https://doi.org/10.1016/B978-0-12-812159-7.00001-1

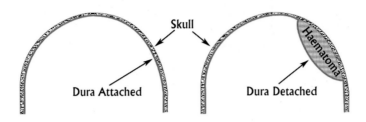

Figure 1.1 Coronal section through the skull.
On the left the dura is attached (gray) and on the right, it has separated from the bone. This opens up an epidural space into which blood can accumulate. For this to happen the dura must be detached in the great majority of cases following a blow to the head which may or may not produce a fracture.

(A)   (B)

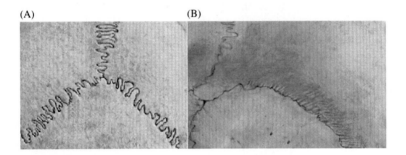

Figure 1.2 Cranial sutures.
(A) Interdigitated sutures ensuring firm attachment between the individual bones.
(B) The joints may be less interdigitated/easier to confuse with a fissure fracture.

2. Meninges

These are the membranes which cover the brain and are divided into three layers. Thus, the outer one is called the dura mater and is attached to the skull and has folds which separate different portions of the brain. Dura means hard and mater means mother as the ancients perceived the membranes enfolding the brain much as a mother cuddles her baby. The brain itself is covered by the pia mater, where pia means soft. It follows the contours of the brain. The intermediate layer is mentioned in the section on cerebrospinal fluid (CSF) later.

3. The Brain

This is the location of awareness, consciousness, intellectual function, emotion, personality, movement, and feeling. To function adequately it needs an adequate blood supply. Its cells cannot regenerate. Its functions may be damaged in the present context if it is shaken about, torn, or if its blood supply is impaired. If it is shaken about it may temporarily lose function. This is concussion.

(A)   (B)

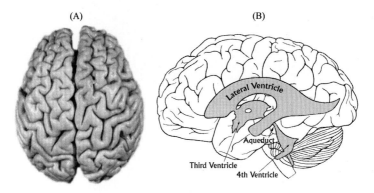

Figure 1.3 The brain.
(A) A cerebrum viewed from earlier after removal of the pia mater and superficial blood vessels.
(B) A diagram of the brain showing the hollow spaces (ventricles) included the interior and indicating the cere-
bellum below and behind. From The History of the Gamma Knife, with permission from Elsevier.[2]

It may otherwise be damaged by tearing cells by direct mechanical damage or following bleeding into its substance. It may also stop functioning if compressed because of loss of blood supply. Fig. 1.3 shows its basic shape.

4. **The Blood Vessels**
   a. On each side of the brain there is a carotid and vertebral artery which feeds blood into an anastomotic arterial circle called the Circle of Willis which supplies the brain. See Fig. 1.4.
   b. The veins from the surface of the brain convey blood into channels called venous sinuses which are unique to the head. They drain the blood away. See Fig. 1.5.

5. **Cerebrospinal Fluid**
   This is a clear fluid produced within the cavities of the brain. It circulates and is absorbed into the venous sinuses. See Fig. 1.6.

6. **Cranial Nerves**
   There are 12 pairs. Some control the movements of the eyes, face, jaws, and mouth. Some receive the ordinary surface sensation from the skin and mouth. Some receive the special senses of vision, hearing, taste, and smell. One pair has multiple functions that include regulating the heart beat and affecting the motility of the gut. They are called the vagi or wanderers and are relevant for the current text.

The acquisition of the information listed earlier occurred in stages with long gaps in between. In principle, most of the early understanding was concerned with anatomy. No real insight into the physiological

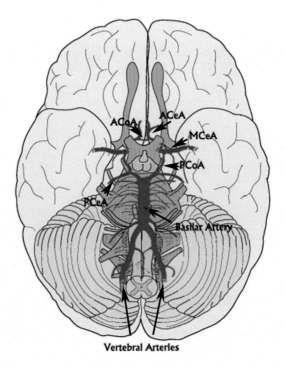

Figure 1.4 Brain arteries.
The course of the main arteries shown in relation to the base of the brain and brainstem.
ACeA, Anterior Cerebral Artery
ACoA, Anterior Communicating Artery
MCeA, Middle Cerebral Artery
PCoA, Posterior Communicating Artery
PCeA, Posterior Cerebral Artery. Adapted with permission from Elsevier Crossman and Neary, Neuroanatomy 3e www.studentconsult.com.[3]

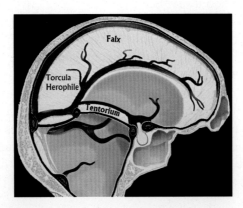

Figure 1.5 Dura and sinuses.
The major folds of the falx between the two hemispheres and the tentorium between the cerebrum and the cerebellum are shown. The venous sinuses are also shown. The torcula as mentioned in Chapter 3, Ancient World—Developing Knowledge. How does it look like a wine press?

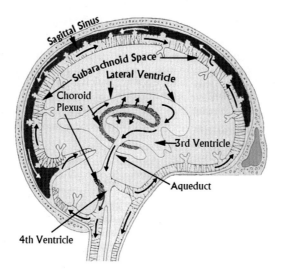

Figure 1.6 CSF circulation.
This figure indicates where the CSF is generated in the ventricles and how it flows out of the fourth ventricle and then upwards and over the brain, in the subarachnoid space to be absorbed in the venous sinuses. The subarachnoid space lies beneath intermediate layer of the meninges, the arachnoid. From The History of the Gamma Knife, with permission from Elsevier.[2]

processes began until the early 18th century. That is not to say that there were no physiological notions. It is just that they were neither correct nor helpful.

The first section of the book will trace the development of anatomical and clinical concepts from the ancient world up to the monograph of intracranial epidural bleeding published by Jacobsen in 1886.[1] It is a measure of the quality of Jacobsen's work that there is little to add on the topic of the clinical manifestations of epidural bleeding, since his seminal publication. It may be emphasized that the evolution of these ideas was a process which by no means unfolded in a straight line.

The second section of the book briefly deals with the clinical expression of epidural bleeding in the modern world. The treatment is either observation or operation and the underlying principles are simple and easy to demonstrate. Even so, even in the 21st century, successful treatment depends on the patient reaching the operating table soon enough. The frustrating reality is that the nature of the clinical condition means that a proportion of patients do not get treated until it is too late. This is the result at least in part of the peculiar pathophysiology of epidural bleeding.

The third section of the book is concerned with this pathophysiology and includes a general consideration of intracranial pressure and cerebral blood flow which is relevant to the understanding of the mechanisms underlying this kind of hemorrhage. This is followed by discussion of experimental findings specifically related to the pathophysiology of intracranial epidural bleeding.[4] This bleed differs from bleeding into any other intracranial space. It occurs into an extracerebral space which, as indicated earlier does not normally exist, but which can open following an appropriate stimulus such as trauma or in very rare cases rapid CSF drainage. This epidural "space" is moreover a cavity in the walls of which are many torn veins which provide a drainage system of great capacity through which epidural blood can escape. The book will close with a consideration of how the excellent modern treatment could possibly avoid the dangers of treatment delayed too long.

REFERENCES

1. Jacobson WHA. On middle meningeal haemorrhage. *Guys Hosp Rep*. 1885/1986;43:147−308.

2. Ganz JC. *The History of the Gamma Knife Amsterdam*. London, New York: Elsevier; 2014:4−5.

3. Crossman AR, Neary D. *Introduction and Overview. Neuroanatomy*. Edinburgh, London, New York: Churchill Livingstone; 2005:9.

4. Ganz JC. *Pathophysiology of Supratentorial Intracranial Epidural Bleeding: An Experimental Study*. Bergen: Universitet i Bergen; 1990.

History of Brain Trauma Management

CHAPTER *2*

Ancient World—Before Brain Anatomy

INTRODUCTION

Illness antedates medicine even affecting the dinosaurs.[1] More oddly, surgery antedates literacy as evidenced by trepanation; as outlined later. The early days of neurosurgery evolved out of superficial abnormalities as there was no facility for classifying, diagnosing let alone treating illnesses deep within the cranium. This means in effect that all early neurosurgery was trauma related. The subject of this book is epidural bleeding which is a relatively uncommon complication of some head injuries. In order to understand how it came to be known and eventually managed it is necessary to understand the background from which such management emerged so the earlier parts of this book will recount the evolution of the management of such trauma. This covers a long period, since epidural hematomas per se began to be noted in the 18th century and it was not until the 19th century that modern understanding of the condition began to emerge. A natural place to start the description of this evolution is the earliest mentions of head injury management in the ancient world.

ANCIENT EGYPT

The first documentation of what would be regarded as modern case reporting on head injuries is the Edwin Smith Papyrus.[2] The papyrus was bought in Luxor in Upper Egypt in 1862. It was bought by Edwin Smith an American who lived in Luxor for 18 years between 1858 and 1876. The seller was Mustafa Agha, a dealer, merchant, and Egyptian Consular Agent in the town. It was an ancient manuscript roll with some outer bits missing. Two months later Mr. Agha sold Smith some remaining fragments glued onto a fake roll. Smith recognized this fraud and also the value of the fragments which he also purchased. His personal reputation has been variously reported. The distinguished Egyptologist Warren Royal Dawson described him as an adventurer,

Intracranial Epidural Bleeding. DOI: https://doi.org/10.1016/B978-0-12-812159-7.00002-3

money lender, antiques collector and dealer, and one who had familiarity with antiques forgery. James Breasted the celebrated American Egyptologist had great respect for Smith's knowledge of the hieractic Egyptian language, even if this was insufficient to translate the document.[2-4]

Smith died in 1906 and his daughter donated the papyrus to the New York Historical Society in the same year. In 1920 the Society asked James Breasted, Director of the Oriental Institute at the University of Chicago, to provide a decent translation. The final edition was published in 1930. Breasted considers that it contained writings from three persons. First, there was the original author who wrote in the language of the Old Kingdom and must have written between 2600 and 2200 BC. Several centuries later a commentator added 69 short explanations or glosses. The ancient language had fallen into desuetude and these glosses were clues to the meaning. Finally, there was a scribe who copied the manuscript. He wrote beautifully, alternating red and black ink without much system. He was careless making errors, he must subsequently cross out. But halfway down he stopped, leaving the rest of the papyrus blank. He is believed to have worked in around 1650 BC. Twenty-seven of the cases described in the document were cases of head trauma. The cases presented were divided into title, examination, diagnosis, and treatment. The whole layout was modern and there is no reference to supernatural influences.[2]

However, while the descriptions of the injuries were comprehensible in modern terms there was no case of an epidural hemorrhage. Nonetheless, there are some other features of ancient practice which bear mention in this context. These relate to clinical observation, anatomical knowledge, and the basis for medical/surgical treatment. Some of these descriptions clearly illustrate essential features related to cranial trauma. A difference between a swelling under a smashed bone and the appearance of cerebral convolutions is noted. The following is written in the description of a fracture is in Case 4 "A gaping wound in the head penetrating to the bone and splitting the skull you should palpate the wound. If you find something disturbing therein under your fingers and he shudders exceedingly while the swelling which is over it protrudes." This is consistent with a description of an open fracture with raised intracranial pressure and external herniation of the dura covering the brain when the patient shudders. In Case 6 on the other hand the description is "rending open the brain of the skull." It goes on to mention an appearance of "corrugations which form in molten copper" which are presumably cerebral convolutions. It also mentions "something therein throbbing and fluttering under your fingers" which is a good description of brain exposed after open trauma together with its pulsations. In this same case, there is a Gloss A which mentions "'Smashing his skull, rending open the brain' (it means) the smash is large, opening to the interior of the skull, (to) the membranes enveloping the brain, so that it breaks open his fluid in the interior of his head" suggesting that cerebrospinal fluid (CSF) and dura were both observed. If that is correct, this case involves the first ever description in history of the brain, dura, and CSF.[2] Thus, the structures observed were described without any indication that their function was understood. In a number of cases, neck stiffness is noted; a common concomitant of blood in the subarachnoid space from any cause including trauma. In others bleeding from the nose and ears is mentioned, indicative of skull base fractures. In Case 8 a spastic paresis of a foot with a contracture is mentioned which is considered due to the head injury.[2] The significance is not taken any further than just to record the phenomenon.

The writer seems also to have had some idea of the severity of the injuries since the cases are classified into I shall treat, I shall contend, and I shall not treat. Why did the Egyptian physician not understand what he was looking at? Well first, he had no means of knowing the anatomy of the central nervous system because the technology for discovering this would not be available for many centuries. There is however more. The Ancient Egyptian did not seem to consider the brain as important. During mummification it was removed through the nose via a hole in the ethmoid using a hook.[5,4] The dorsum sella (a little bony wall behind the pituitary gland) was often fractured and left lying in the back of the cranium. The brain was not kept with the body like some other organs especially the heart as it was clearly deemed not to be important for the afterlife.[6] Its precise role in this civilization is simply not clear. Other ancient societies also considered the brain to be unimportant. The ancient Chinese believed the brain to be the marrow of the skull. So the writer of the Edwin Smith Papyrus would not consider if the brain could control movement.[7] While Egyptian physicians must have known as indeed primitive man had known before him that a head injury could paralyse or even kill,[8] nonetheless, there was no connection made between that knowledge and the presumption that the brain had any useful function; a reflection on how humans process observations.

A NOTE ON SCIENTIFIC OBSERVATION

It is worth digressing for a moment on the thoughts of Claude Bernard on the nature of observation which he carefully analyzed in the introductory chapter of his monograph "An Introduction to the Study of Experimental Medicine." He is at pains to point out that observation is not just a passive process. He takes the case of observing planets. He suggests that if by chance an unexpected irregular movement of a planet is observed this is passive and chance. However, if following this observation a search is started for the cause of the irregularity, what is found during that search is an active process.[9] Thus, it is feasible to believe that many astronomers could observe the irregular movements of a given planet and simply write it off as an oddity. What is observed is also limited by the instruments available for the observation which in ancient Egypt were the observer's physical senses. It takes analysis and the application of logic to start seeking for a cause. In the case of brain function and the Egyptian physician that step was not taken due to another probable limiting factor; the concepts of disease current at that time. This requires brief consideration.

EVOLUTION OF CONCEPTS OF DISEASES

In a single volume history of medicine Roy Porter emphasizes how the patterns of disease evolved following the evolution of homo sapiens.[10] He points out that as long as humans were nomadic hunter foragers, most of their disorders would be trauma and diet related, thus emphasizing the primacy of trauma in the development of human suffering. Later on, when shortage of resources compelled humans to band together in groups and start farming, the pattern changed. At this stage the numbers gathered together in one place provided a reservoir for the transmission of infectious diseases. Moreover, the animals required for husbandry provided another reservoir of infectious diseases. These infections could be vicious, violent, and incomprehensible. Thus, it is not surprising that in most if not all primitive societies the initial concept of disease consists of the interference by a mystical, magical, or religious outside influence. Demons, spells, and the malign influence of the dead were all blamed for making a person sick.[5] The Ebers Papyrus sold to Georg Ebers in 1872 was finally available in a reliable English translation by Paul Ghalioungi in 1987.[4] This papyrus begins with three spells or incantations. The incantation reads "'Here is the great remedy. Come you who expels evil things in this my stomach and drives them out from these my limbs! Horus and Seth have been conducted to the big palace at Heliopolis, where they consulted over the connection between Seth's testicles with Horus, and Horus shall get well like one who is on earth. He who drinks this shall be cured like these gods who are above'... These words should be said when drinking a remedy. Really excellent, proven many times!"[11] There you have a characteristic invocation to external supernatural powers to facilitate healing. In China in early times disease was considered to be visited upon people by angry departed souls or by demons.[12] In Ancient India the origin of medicine is magico-religious. In Mesopotamia, mysticism and religion were intimately involved in healing and the Sumerians had developed a system of divination.[13] In ancient India, Hindu medicine is called Ayurvedic medicine. The roots of this medical tradition go back to Vedic hymns which contain magico-religious doctrines of medicine. They go as far back as 2000 BC.[14] And of course, the most familiar examples of healing by magic are Christ's healing miracles in the New Testament of the Bible. Without going into more detail, it is fair to assert that this preliminary phase of medical development was widespread if not in fact universal. It was primitive man's simplest response to familiar phenomena of daily life for which he did not have and could not have a rational explanation.

ANCIENT GREECE—FROM RELIGION TO REASON

This approach was also present in ancient Greece which is the place where systematic rational medicine was to have its origins. In Greece, the leading religious system of medical management was centered round the cult of Asklepios. The origins of this cult are like a mini tour through classical Greece. The tale begins on Delos with the god Apollo, the god of the sun but also the god of healing. He is reputed to have been born in a grotto on Delos. This island was an important commercial center of the Delian League. Yet even in Ovid's time (43 BC–AD 17) it had shrunk to "Delos the desolate, where once men prayed".[5] Today it is a gorgeous little uninhabited island surround by royal blue sea. It may only be approached in a small boat which seems to ride high in the seas and gives a sense of how uncomfortable a long journey across the eastern Mediterranean must have been in ancient times.

Apollo moved from Delos to Delphi, where the ruins of his magnificent temple may be found today. The main temple is above the road next to the small spring which irrigates the olive groves in the valley down the side of Mount Parnassos. The atmosphere is awe inspiring, particularly in the late afternoon and early evening. Sleeping out under the full moon next to the Tholos (part of a site dedicated to Apollo's sister Athena) nearly a kilometer from the main temple is by contrast a very peaceful and tension removing experience. Let it be said that the ancient Greeks had an uncanny knack of finding the best sites for temples and important buildings.

The story was that Apollo fathered Asklepios, but that the mother was unfaithful and Apollo's sister Artemis killed her while Apollo rescued the baby from the body and gave him to Chiron the centaur to train.[5] Asklepios was according to the story killed by Zeus, the reasons given being varied but disturbing the accepted social order was high on the list of possible causes; perhaps for being cheeky enough to bring people back from the dead. After his death Asklepios was deified. During his life, he had three daughters, Meditrine, Hygeia, and Panacea. Two of these names have passed into every day medical language. The system of healing that Asklepios shall have designed was centered round a temple known naturally enough as an Asklepion; plural Asklepia. The patient would come to the temple and sleep overnight. The process was called incubation from a Greek word meaning to lie down. The next morning the patient would recount any dreams he or she had to the priest, who would interpret them. There were snakes kept at the asklepia which might lick the patient during the night. The snakes concerned were Elaphe longissima longissima, a tree snake and the only European constrictor. It is harmless to humans.

The staff of Asklepios with its single snake should not be considered to have a relationship with caduceus, the staff with two snakes used so often as a medical symbol (see Fig. 2.1). This derived from Hermes (the messenger of the gods) and its association with medicine is not clear. It is a modern innovation within the last 100−200 years.

The priest advised about treatment, which was supposed to be prescribed by the god. The patient might leave a votive offering, often a

Figure 2.1 Statue of Asklepios in Epidauros. Note the single snake round his staff. From Shutterstock.com.

representation in stone of the affected part. The most famous asklepion is in Epidauros in the southern part of Greece, roughly 90 miles from Athens. The buildings are beautiful and the open-air amphitheater by some miracle of acoustics permits a low voice on the stage to be clearly audible in the highest row. It is a place of great atmosphere, not like Delphi but still impressive in its own particular way. This would of course lend to the effectiveness of any treatment. Asklepia continued to be constructed for many centuries. Asklepios is supposed to have been born around 1250 BC. As will be mentioned later new asklepia were being built over 900 years later.[15]

TREPANATION

Discovery

For those trying to trace a logical thread through the development of surgical understanding, trepanation is a monumental anomaly. It is dramatic but mysterious. Its practice seems to lie outside the study of medicine outlined earlier. The attraction for this form of surgery antedates rational medicine, it antedates not only primitive religious medicine but also the development of writing. It makes one ask the question "what is so fascinating about surgery?" The reply has to include the drama. However, the lure of the dramatic should not lead us into an uncritical assessment of this phenomenon. A serious fairly recent review cautions us against being too free with our interpretation of trepanation.[16] It suggests care must be taken about treating any hole in an ancient skull as the result of an operation. There should be preferably signs of healing or signs of surgical opening. Other postmortem agencies can make holes in bone. With advantage this may be considered in more depth.

The first trepanation was described by Paul Broca, a celebrated French neurologist. In 1865, Dr. Prunières, a general practitioner from Marvejols, had found a skull with a hole in it. He believed the hole was related to the postmortem use of the skull as a drinking vessel. However, Broca in 1876 pointed out the smooth edges of the opening, indicating healing.[5] Subsequent archeological finds have shown that the practice was widespread. The oldest operation was Mesolithic around 10,000 BC from North Africa and there are skulls from near Jericho from 8000 to 6500 BC. However, the majority are from the Neolithic area. Trepanned skulls have been found in so diverse places such as Peru,[5,16] Columbia,[16] Mexico,[16] United States,[16] Canada,[16] France,[5,17] Austria,[5,17] Poland,[5] Russia,[5,17] Germany,[5,17] Spain,[5] Great Britain,[5,17] Italy,[17] Denmark,[17] Middle East, Africa,[5] Asia, and Pacific islands.[5] Thus, it would seem to have developed independently in a great number of cultures. It is of interest that there is evidence that trepanation was undertaken in ancient Egypt, the home of the first description of the brain and dura but that it must have been very infrequent.[2]

Indications
There is clear evidence that a major indication for trepanation has been trauma. This is shown particularly in Danish and Peruvian[16] skulls with a preponderance of injuries in men often in the left temporal region; the most likely location to be hit by a club wielded from in front by a right handed foe. These skulls also often have a fracture line associated with the trepanation. The anatomical distribution of trepanation openings is however varied and they are found in any location

including over the sagittal sinus. A great number of them show signs of healing with regrowth at a smooth edge. To gain insight into the reasons for performing these operations anthropologists have investigated contemporary primitive tribes who have practiced trepanation within modern times. They have found in the Atlas Mountains of Algeria that the main indication was trauma. However, in East Africa there were indications that trepanation was used for various conditions related to the head including headache, epilepsy, and vertigo. Moreover, within East Africa, there were fairly marked regional variations in practice over quite short distances. In New Ireland, an island near Papua New Guinea, it has been found that many men had trepanation in their youth as an aid to longevity.[5] This finding of nontraumatic indications in recent times taken together with the finding that not all prehistoric skulls show signs of trauma suggests that there have always been magical mystical indications for trepanation. However, it is not possible to be precise about it. One other indication could be frontal sinusitis. This has been described in a 5500-year old skull.[16] Most peculiarly, in the Atlas Mountains in Algeria there was the oddest of indications recorded in recent times. This indication among other things showed how safe the people considered the operation to be. Women would undergo trepanation to provide evidence that their husbands assaulted them when they wanted grounds for divorce.[18]

Technique

Despite the fact that trepanation was developed long before literacy, there is clear evidence that different surgeons had different techniques, thus indicating a fundamental characteristic of a craft based on manual skill. The skulls have been scraped, cut, chiseled, and drilled.[16] The anesthesia used in Neolithic times remains unknown but in East Africa in modern times it is reported that the only pain involved was related to cutting the skin.[16] The manipulation of the bone was not painful. This is in keeping with the experience of any modern neurosurgeon who has drilled the skull under local anesthesia. The skin and pericranium require anesthesia. Otherwise the only other painful structure is the chance placement of a meningeal artery under a bone opening. Drilling the bone itself is not a painful procedure. In visiting the Algerian operators, a British anthropologist Hilton-Simpson[5,18] recorded that on no account was the dura to be injured during trepanation. Moreover, the sutures were left untouched. They were considered to be the patient's destiny written by the hand of Allah.

Present Day

Even today there are people who believe that making a hole in their own head would serve some useful purpose, claiming that it would increase the blood flow to the brain. There are circumstances in cases of injury and stroke where there is raised pressure in the head and removing a part of the cranium may help to compensate for the raised pressure and improve the circulation to parts of the brain that are not irrevocably damaged. However, this is a procedure demanding a massive bone removal and opening of the dura mater beneath. Moreover, even this draconian procedure would have no benefit for an uninjured brain, since within wide limits the volume of blood delivered to the brain is kept constant by an inherent mechanism called autoregulation (see Chapter 11, The Lucid Interval). It is a matter of frustration and sadness that such practices are undertaken. It is a matter of grave misgiving that anyone engaged in such idiocy should receive any money from either private or public investors.

CONCLUSIONS

In a primitive society, everyday phenomena from the weather to health have no ready rational explanation. Supernatural interpretation is seemingly universal and unavoidable. This is the best that people can do on the basis of limited data. A byproduct of this phenomenon is the growth of a priestly class who interpret the supernatural and whose income and influence depend on the maintenance of accepted doctrine.

It is a matter of wonder that humanity, against the opposition of the priests, still queries the nature of existence and in the end after a long battle wins through. The existence of trepanation remains a somewhat inexplicable phenomenon. However, in a book on epidural bleeding, any knowledge of the dura is relevant and the Edwin Smith Papyrus dating perhaps from 2500 BC indicated awareness of the brain and dura, though without any indication that their purpose and functions were understood.

There was what might be called a hunter's knowledge of the injured head. There was awareness of damage to the encasing bone which if severe enough could reveal a pulsating structure within. It is possible that CSF was observed but not understood. There seems to have been no understanding of what lay within the dura. There was however an instinctive effort to classify the severity of injury. This would be passive observation without any analysis of underlying mechanisms.

REFERENCES

1. Molnar RE. Theropod paleopathology: a literature review. In: Tabnke DH, Carpenter K, eds. *Mesozoic Vertebrate Life*. Bloomington, IN: Indiana University Press; 2001:337–363.

2. Breasted JH. The Edwin Smith Surgical Papyrus. In: Wilkins RH, ed. *Neurosurgical Classics*. Rolling Meadows, IL: AANS Publications; 1992:1–5.

3. Majno G. *The Healing Hand*. Cambridge, MA: Harvard University Press; 1965.

4. Nunn JF. *Ancient Egyptian Medicine*. London: British Museum Press; 2000.

5. Guthrie D. *A History of Medicine*. London: Thomas Nelson and Sons Ltd; 1960.

6. Finger S. The brain in antiquity. In: Finger S, ed. *Origins of Neuroscience: A History of Explorations into Brain Function*. Oxford: Oxford University Press; 1994:6–10.

7. Finger S. The brain in antiquity. In: Finger S, ed. *Origins of Neuroscience: A History of Explorations into Brain Function*. Oxford: Oxford University Press; 1994:12.

8. Finger S. The brain in antiquity. In: Finger S, ed. *Origins of Neuroscience: A History of Explorations into Brain Function*. Oxford: Oxford University Press; 1994:3–4.

9. Bernard C. *An Introduction to the Study of Experimental Medicine*. New York: Dover Publications Inc; 1957. 5–26 p.

10. Porter R. The roots of medicine. In: Porter R, ed. *The Greatest Benefit to Mankind*. London: Harper Collins; 1997:14–44.

11. Glendening L. *The Egyptian Papyri. Source Book of Medical History*. New York: Dover Publications Inc; 1960:4.

12. Unschuld P. History of Chinese medicine. In: Kiple KF, ed. *The Cambridge World History of Human Disease*. Cambridge: Cambridge University Press; 1995:22–27.

13. Singer C, Underwood EA. *A Short History of Medicine*. 2nd ed Oxford: Oxford University Press; 1963.

14. Gallagher N. Islamic and Indian medicine. In: Kiple KF, ed. *The Cambridge World History of Human Disease*. Cambridge: Cambridge University Press; 1995:30.

15. Singer C, Underwood E. *A Short History of Medicine*. 2nd ed Oxford: Oxford University Press; 1963.

16. Saul FP, Saul JM. Trepanation: old world and new world. In: Greenbalt SH, ed. *A History of Neurosurgery: In Its Scientific and Professional Contexts*. Park Ridge, IL: The American Association of Neurological Surgeons; 1997:29–35.

17. Ortner DJ, Theobald G. Diseases in the pre-Roman world. In: Kiple KF, ed. *The Cambridge World History of Human Disease*. Cambridge: Cambridge University Press; 1995:248–249.

18. Majno G. *Prelude. The Healing Hand*. Cambridge, MA: Harvard University Press; 1965:1–29.

CHAPTER 3

Ancient World—Developing Knowledge

INTRODUCTION

Epidural hematomas occur between the dura and the inner surface of the cranium. They compress the underlying brain through the dura. This means that in considering the medical history that relates to or could relate to such hemorrhages it is necessary to focus attention on specific areas of knowledge. These include the structure and functions and inter-relationships of the cranium and its coverings, the meninges, and the brain, including its blood supply and its all-enveloping cerebrospinal fluid. This chapter is concerned with dawning knowledge of the anatomy of the skull and its contents and how that knowledge affected clinical practice. First there is brief précis of the beginnings of anatomy and physiology. This is followed by an account of the relevant contributions of the three major pioneers, Hippocrates, Celsus, and Galen. This chapter ends with a summary of early understanding of neurological disease.

BASIC SCIENCE

Anatomy

It follows from the previous chapter, that prior to the Ancient Greeks, interest in and observation of animal anatomy had been a passive practical matter and not the object for academic analysis. On the other hand, dissection to quote one of the leading historian specializing in these times, Sir Geoffrey Lloyd, "is always an experimental procedure in the weak sense in that it involves not simply direct observation, but 'observations prvoqueés' that is, a piece of research deliberately undertaken to discover facts".[1] This is in keeping with Bernard's notions of active observation. The first sign of such activity is usually thought to be the work of Alcmaeon of Kroton (born c. 510 BC) (see Fig. 3.1).[2] He has been accredited with dissection of the optic nerve but modern expertise suggests that this is not the case. It was rather that he removed an eye and saw the optic nerve disappearing in the direction

Intracranial Epidural Bleeding. DOI: https://doi.org/10.1016/B978-0-12-812159-7.00003-5

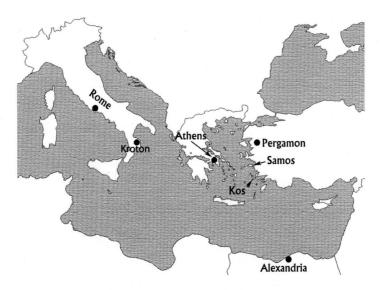

Figure 3.1 Outline map of the Mediterranean showing the locations of important scientists. Samos is where Pythagoras was born and Kroton in Italy where he worked. Alcmaeon was also from Kroton. Hippocrates was born on Kos. Athens is the capital of Greece and Rome is the capital of the Roman Empire. Galen was born in Pergamon and worked in Rome.

of the brain. Even so, while this is not detailed dissection, it was still a formal attempt to acquire, however limited, a piece of anatomical knowledge. The first extant text on cranial injury was the work of Hippocrates from Kos (460−370 BC). It is generally known even from classical times that the Hippocratic writings were not the work of a single person but rather of a group of like-minded individuals. This should be born in mind but in the context of this chapter, the distinction is not important and for simplicity, reference to the group will use the term Hippocrates alone.

Physiology

The ideas on which concepts of physiology in Ancient Greece were based arose out of the intellectual traditions of the time. There was a belief that a balance between the forces of nature and humanity would be expressed in terms of healthy living with a balanced diet and levels of activity. It is encapsulated in the albeit Latin saying "mens sana in corpore sano." As time passed various attempts were made to impose system on the sometimes nebulous and diverse observations and concepts that were current at the time. These systems would have to be imposed on a mixture of magical, mystical, and religious concepts and beliefs. They were derived out of a fascination with numbers.

Pythagoras (c. 570−475 BC) was the person who was truly at the beginning of the process which led to an accepted synthesis to be used for many centuries. He was a great philosopher born on Samos, a Greek island close to Turkey. Indeed, the word philosopher arose out of one of his sayings. He was asked once if he were wise and replied "No I am a lover of wisdom"[1,3] (philo-sophos in Greek). Pythagoras and his followers were interested in, or more accurately seemingly obsessed with numbers. Their most sacred number was 10. Thus, since the series of numbers $1 + 2 + 3 + 4 = 10$, this series is of prime importance.[4−6] Four represented completion. It followed that there should be a number of elements that made up everything and that this number was four.[6] Thus, Pythagoras influenced the number of elements required to make matter. Empedocles (c. 490−430 BC) of Agrigentum finally came up with the cosmogenic theory of four classical elements which captured the minds of those who came after.[7,8] Aristotle would define an element as the primary constituent in something—be it object, speech, or a geometrical proof—which is indivisible into any other kind of thing.[6] This is basically the meaning we use today and the Latin "Elementum" is defined as "the simplest component of a complex substance". Empedocles' elements were air, water, earth, and fire. It is interesting to note that his concepts are pleasingly modern, involving notions equivalent to conservation of mass and energy. It was his contention that the elements were present in unchanging amounts. Their interactions were governed by two forces "love" and "strife." They combined into different forms and then split up again, producing all matter and states of matter. However, the underlying elements remained constant. Only the different forms of their various combinations altered, giving the illusion of changes in the matter; which is notwithstanding conserved. It is interesting though inexplicable why this particular system for explaining the laws of nature should be accepted: but accepted it was and held sway for nearly 2000 years. It is appropriate to remember that the concept of elements was not limited to western civilization. The Chinese had 五行 or the "Wu Xing" meaning five elements. These were similar to Empedocles' but not identical. The Chinese elements were air, water, earth, fire, and metal.

From the four elements, a system of general physiology developed, which is usually credited to Hippocrates. The qualities of hot, cold, wet, and dry are associated with the four elements, thus fire is hot and

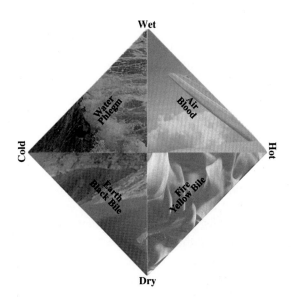

Figure 3.2 Illustration of the relationship between the qualities of the elements and the physiological humors. See text.

dry, water is cold and wet, earth is cold and dry, and air is hot and wet. The humors are Blood, Yellow Bile, Black Bile, and Phlegm. Blood is hot and wet. Yellow Bile is hot and dry. Black Bile is dry and cold and Phlegm is cold and wet as illustrated in Fig. 3.2 and their relationship to the four elements and qualities is also shown.

The concept behind the humors was that they should be in balance in a healthy patient. This was a state of eucrasia = eu (normal) + krasis (mingling). Excess of one humor led to imbalance and sickness. The word was dyscrasia from dys (abnormal) + krasis (mingling). The word dyscrasia has persisted into modern times being used only a few years ago as part of the classification of blood disorders. In addition to being associated with certain patterns of behavior the humors came to be associated with certain organs of the body. With this unfamiliar background, out of the way the nature of clinical advances in ancient Greece may be reviewed.

CLINICAL SCIENCE—HIPPOCRATES

Background
Nothing is known about Hippocrates except from his writings and the comments of others. There is no genuine biographical material.

However, he worked on battlefields and must have traveled extensively round the eastern Aegean, which would involve sea travel, which in sail boats in that part of the Mediterranean is not for the faint hearted. He must have been possessed of considerable strength of mind since his teachings were quite contrary to those of Asklepios. While there is no evidence, one can easily imagine that there were many testy arguments within the paternal home between this upstart youngster and a father and maybe grandfather who were both priests of Asklepios. No matter family traditions, Hippocrates departed from religion and mystery and with his precise observations taught that disease had a natural and not a supernatural origin. His contribution to medicine remains unsurpassed. Lloyd mentions that there is scant evidence of dissection in the works of Hippocrates, with the exception of "On the heart."[1] Yet it will be seen that he had considerable knowledge of anatomy and organ function relating to cranial injury. How this could this have been acquired without dissection will become clearer.

Concerning the Brain

Hippocrates (c. 460 BC—370 BC) was convinced of the importance of the brain. This he stated as follows in "On the Sacred Disease." "Men ought to know that from nothing else but thence (from the brain) come joys, delights, laughter and sports, and sorrows, griefs, despondency, and lamentations. And by this, in an special manner, we acquire wisdom and knowledge, and see and hear, and know what are foul and what are fair, what are bad and what are good, what are sweet and what unsavoury .... And by the same organ we become mad and delirious, and fears and terrors assail us, some by night, and some by day, and dreams and untimely wanderings, and cares that are not suitable, and ignorance of present circumstances, desuetude, and unskillfulness. All these things we endure from the brain, when it is not healthy."[9] Despite Aristotle's contention that the heart was the center of the personality, all the physicians considered later were convinced this function was the property of the brain.

Injuries of the Cranial Bones

"On Injuries of the Head"—is the first still existing neurosurgical text. Hippocrates mentioned that skull thickness varies in different locations and the sutures were described. In particular, he described the sagittal, coronal, and lamboid sutures in a variety of ways. On the other hand, Hippocrates did not allocate a name to any of the sutures. He considered the bone was thinnest at the upper frontal region which has the most brain beneath it, making it a vulnerable location compared with the back of the head. The temporal region was also perceived to have thin bone and a large vein crossing it. It is not stated which vessel this might be, but it is tempting to think that it was a superficial temporal vessel. His assumption of the importance of the brain is expressed by the notion that injury to the upper frontal region would more likely lead to death because of the thin bone and larger portion of the brain in that region. He also made repeated mention of the dura throughout the monograph. In addition, he mentioned that the brain is covered by two membranes. In "Places in Man" it is stated "Of the brain, there are two membranes, the fine one cannot be healed, when once injured."[10] Thus, while he did not dissect he had considerable anatomical knowledge of the coverings of the cranium and its contents, presumably acquired during the management of head injuries not least on the battlefield. It is knowledge based not on systematic explorative analysis but on practical experience.

He described five types of fracture; fissure surrounded by contusion, contusion without fracture, depressed fracture, hedra with or without contusion or fracture, center-coup injury.

Etiology

Hippocrates went into some detail describing different trauma mechanisms and their relationship to the bony changes produced. The nature of the implement causing the wound and how it struck the head should be known. This would suggest the likely injuries to be found. For example, a glancing blow is less likely to produce a fracture than a blow vertical to the surface of the bone. Round, globular smooth, blunt, heavy, and hard instruments are more likely to produce fractures and depressions. The injuries so produced will be more likely associated with crushed scalp, suppuration, and slow healing. In contrast, oblong weapons which are slender sharp and light penetrate the scalp rather than bruise it and while they may produce a hedra or cut in the bone they do not cause contusions, fractures, or depressions. Hedras are considered later.

Investigation

There is no advice on examining an injury if the skin is intact but in the presence of any skin defect the wound should be extended to ensure adequate examination of the bone. This examination included scraping the bone with a raspatory to follow any injury into the depths. If there was doubt about the presence of a fracture or contusion, a black die may be applied to the wound and kept in place with a poultice of flour and vinegar. When extension of a skin laceration was needed for adequate revelation of the underlying wound, the extended wound should be filled with a large dressing called a tent, which would expand the wound by the next day when the raspatory was again used if a fissure were revealed stained black with die. Moreover, contused bone would be revealed having imbibed the die. It should be noted that part of the aim of this procedure was to distinguish between suture and fissure. These procedures were the equivalent of today's imaging studies. They were repeated in surgical texts for nigh on two millennia after Hippocrates' death. If there is no adequate laceration, the state of the underlying bone must still be considered. In some cases, the injury would be examined with a probe to ascertain if any bone was depressed or if there was a significant fracture.

Management

Cranial wounds should be kept dry and dressings should not be placed in them as treatment except in the frontal region which is bare of hair. The only other indication for using subcutaneous dressings or tents was to facilitate the clarification of the extent of an injury as mentioned earlier. Crushed skin would tend to suppurate and slough away. It was desirable to expedite this suppuration so that the integuments become clean and not a source of suppuration to damage the bone. However, it is not precisely stated that such tissue should be excised although that seems to be the implication. If the dura was exposed the same principles apply to ensure that the membrane did not become damaged because of surrounding suppuration.

The indications for trepanation were fissure, contusion, and hedra. It is not clear why a fissure required trepanation, though maybe there was concern that there was material underneath which should be allowed to escape. The text is not clear on this point. However, the issue of contusions and hedra require clarification.

Clarifications

Before proceeding to the matter of trepanation itself it is necessary to consider notions which are unfamiliar to a modern reader. First, there is the term contusion. Today we think of a contusion as the swelling which occurs after an injury in which there is leakage of blood from broken, usually small blood vessels into the tissues. Such a concept would not have been available until the 19th century so Hippocrates would have meant something else but it is not clear exactly what. The concept was still active in the 18th century when Percival Pott mentions it repeatedly in his book on head injuries.[11] Yet it is not a phenomenon familiar to surgeons today. Admittedly, the term bony contusion is to be found on MEDLINE but it refers almost exclusively to magnetic resonance imaging findings mostly around the knee joint. It is not a feature of cranial injuries. The nature of the lesions to which Hippocrates refers must necessarily remain obscure. However, it would appear to be a genuine phenomenon in that in open wounds on the battlefields of ancient Greece something happened to the bones which enabled them to absorb black ink inserted into a wound, thus distinguishing them from surrounding bone.

Then there is the matter of fracture depth. The notion that fractures can affect only the outer table but spare the inner is consistent from Hippocrates onwards for many centuries. This is not something that we see in modern practice. It is not mentioned once in George du Boulay's 372 page tome solely devoted to plain Skull X-rays.[12] However, O'Halloran as late as in the 18th century recalled a case where a man had been struck on the head by a sword and where the injury was limited to the outer table.[13] Since this trauma mechanism is no longer current, and since so many authors have described fractures limited to the outer table it seems reasonable to suggest that such injuries really existed when the head was struck with a sharp instrument such as a sword. This makes the scraping bone to assess fracture depth sensible. However, this does not necessarily apply to the so called hedra. As suggested elsewhere, it seems likely that a hedra was in reality a focal solitary depressed fracture (see Fig. 3.3).[14] The term appears

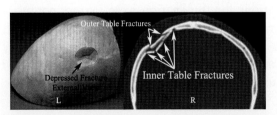

Figure 3.3 A solitary depressed fracture may appear to be limited to the outer table of the skull (L) but the CT (R) shows how misleading this exterior appearance can be with the marked fracturing of the inner table invisible from the exterior. The image on the left is from Wikipedia.

in two current English translations of Hippocrates' text, published in the Loeb Library; one by Adams and one by Withington.[15,16] It should be remembered that the Adams translation of Hippocrates monograph on head injuries was made before X-rays had been discovered and the Withington translation in 1928 at a time when quality skull X-rays would not have been universally available. Thus, these translations are made at a time when modern understanding had not yet been achieved. It seems reasonable to consider that while some fissures could be limited to the outer table, a dent in the skull following a blow with a sword or similar sharp object would necessarily involve both tables, making the injury a focal depressed fracture (see Fig. 3.3). This being so trepanation would be sensible. Hippocrates himself gave a clue to support this argument in respect of the management of loose fragments. He states "A piece of bone that must separate from the rest of the bone, in consequence of a wound in the head, either from the indentation (hedra) of a blow in the bone, or from the bone being otherwise denuded for a long time, separates mostly by becoming exsanguous." If a hedra was merely a dent, perhaps limited to the outer table there is no reason for loose fragments to develop.

Return to Management

Hippocrates considered fissures and contusions should be trepanned and a hedra associated with a fissure or contusion should also be trepanned. An isolated hedra did not require trepanation. He further stated that depressed fractures should not be treated with a trepan since in his experience the depressed fragments would regain their position spontaneously given time. His advice to avoid trepanation seems odd at first and requires further consideration. His description of a depressed fracture is in fact that of a widespread comminuted fracture. For injuries of this kind his advice is sound and there is reason to believe this is a correct interpretation because the precise wording is "Such pieces of bone as are depressed from their natural position, either being broken off or chopped off to a considerable extent, are attended with less danger, provided the membrane be safe; and bones which are broken by numerous and broader fractures are still less dangerous and more easily extracted."[4] This sort of injury is also consistent with the battlefield and there were over 20 battles in Greece during his lifetime. It should also perhaps be remembered that one of his sayings is said to have been either "War is the only proper school for the surgeon" or "He who wishes to be a surgeon should go to war." In the name of accuracy, it must be mentioned that the actual text is somewhat different. This quotation comes from an essay called "De Medico" which is part of the Hippocratic corpus. It is to be found in Section 14 of the essay and deals with military surgery. It states in that context that "Therefore, he who wishes to operate (practice surgery) must serve in the army and follow foreign troops (away from home). This way he can be trained in this matter."[20] This translation was kindly provided by Dr. Maria Pantelia, Director of the Thesaurus Linguae Graecae. While not having the force of the above quoted aphorisms, this text does indicate the importance to Hippocrates and/ or his school of battle surgery in the context of head injuries.

Technique

Hippocrates was very precise as to the correct technique of trepanation. Trepanation should be an interrupted process where the instrument was repeatedly removed from the skull and cooled. The process should be accompanied by sounding so that the instrument did not saw all the way through and damage the underlying dura. When a trepanation was almost complete and so little bone remained to be sawed that it had become mobile the process should cease and the bone would fall out of itself. Hippocrates did not state the nature of his trephine. He does however mention in one place the word "serrated." It is believed that most likely he used an instrument otherwise called a modiolus shown in Fig. 3.4, the handle of which was rotated between the palms of the hands. How a downward force was applied is less clear. There were other ways of operating using bow strings but the basics were still the same with two hands rotating the shaft. This would be the basic design for centuries to come.

Clinical Picture

Hippocrates noted several cerebral symptoms following head injury but only in passing. He mentioned that if a patient had dimness of vision, vertigo, stupor, or fell to the ground that was additional evidence as to the severity of an injury and a supplementary indication that every effort, including the use of the jet-black ointment should be used to identify the fracture. Thus, while he seemed aware of the brain there is no specific attempt to attribute these symptoms to brain injury. His effort is directed toward treating the bone.

Figure 3.4 A modiolus (see text). The central pin is to steady the instrument until a circular groove in the bone is produced, at which point the pin is removed.

Extra Findings
Epidural Bleeding?

The bones of children are thinner and softer and it is in this context that Hippocrates would seem to provide the first description of an epidural collection of blood. He referred repeatedly to the dura which he called the membrane or meninx. He was concerned about it becoming unhealthy if exposed and allowed to become moist. He was concerned about its perforation either by fracture or trepanation and advised ways in which the latter could be avoided. The relevant passage is translated by Adams as follows. "But if the bone is laid bare of flesh, one must attend and try to find out, what even is not obvious to the sight, and discover whether the bone be broken and contused, or only contused; and if, when there is an indentation in the bone, whether contusion, or fracture, or both be joined to it; and if the bone has sustained any of these injuries, we must give issue to the blood by perforating the bone with a small trepan." Withington's translation is as follows. "But if the bone is denuded of flesh you should devote your intelligence to trying to distinguish a thing which cannot be known by inspection—whether there is fracture and contusion of the skull or only contusion, and whether, if there is a weapon hedra, it is accompanied by contusion or fracture, or both of these. If the bone is injured in any of these ways, let blood by perforating with a small trepan, keeping a look-out at short intervals, for in young subjects the skull is thinner and more on the surface then in older persons." Could this be the very first mention in the literature of epidural bleeding? To illuminate this matter, the help of Sir Geoffrey Lloyd was once again sought. He replied that "The Greek word that the Loeb translates 'let' and Adams 'give issue' is apheinai which means 'let out'." Thus, it would seem there are grounds for believing Hippocrates would consider opening the skull to remove an epidural collection of blood in children. This would give him priority in mentioning epidural bleeding.

Lateralization

It is a commonplace today that injury on one side of the brain produces loss of function on the opposite side of the body. This is distinct from what is called cerebral localization which describes what part of a given cerebral hemisphere is the location of specific functions. Accurate definition of localization would have to wait until the 19th century. However, there was a growing awareness over the centuries of one side of the brain affecting the opposite side of the body. This growing awareness was however a slow and halting process.

Hippocrates was aware of lateralization in two contexts. He specified that incisions should be avoided in the temporal region because of a "vein." He stated convulsions occur if a laceration was made in this region only they would be on the opposite side. Later on, writing about lethal injuries he mentioned that some patients developed convulsions on the opposite side from the trauma. He also mentioned they can become apoplectic or as we would say hemiplegic. Only he does not localize the paralysis only the epilepsy. Even so he was an early observer of contralateral neurological disturbance even if he did not seem to understand that the brain was responsible.

CONCLUSION

It was stated at the beginning of this chapter that attention must be focused on scalp, cranium, meninges, and brain. Hippocrates made significant contributions for all four topics.

1. The scalp should be kept dry to avoid suppuration which could damage the underlying bone.
2. He described a variety of bony injuries and how to manage them including indications for and the proper technique of trepanation.
3. He probably was the first to describe epidural bleeding in children.

4. He was aware that the brain was the seat of emotional and intellectual function.
5. He was aware of contralateral epilepsy after certain injuries. He also noted contralateral hemiplegias without considering the brain might be responsible.

CLINICAL SCIENCE—CELSUS

Background

Aulus Cornelius Celsus (c. 25 BC−AD 50) was a Roman encyclopedist who had a more than passing interest in and knowledge of medical matters, including cranial surgery. He lived through the early days of the Roman Empire through the reigns of Augustus, Tiberius, Caligula, and Claudius. So, he must have been familiar with tyranny and terror. On the other hand, living in Rome he would not have had access to battlefields and the trauma he would have managed would have been civilian. While he is presented as an encyclopedist, most Roman heads of household were expected to treat the simpler ailments from which members of the household suffered. Based on his book "De Medicina" his medical talents were way better than suggested by the casual responsibility of a head of household. It was he who wrote "Now the signs of an inflammation are four: redness and swelling with heat and pain."[17] This is not the contribution of a mere amateur.

He was also much concerned with the ethics and humanity of the surgeon. He stated. "Now a surgeon should be youthful or at any rate nearer youth than age; with a strong and steady hand which never trembles, and ready to use the left hand as well as the right; with vision sharp and clear, and spirit undaunted; filled with pity, so that he wishes to cure his patient, yet is not moved by his cries, to go too fast, or cut less than is necessary; but he does everything just as if the cries of pain cause him no emotion."[17]

Classification

There are two sorts of cranial injury mentioned in De Medicina; fissures and depressed fractures.

Etiology

Celsus like Hippocrates taught that the nature of the instrument of trauma and the blow should be assessed. However, his language suggests that he placed more weight on the examination of the wound than on the history report. The phrase he used after specifying questions aimed at determining the details of the trauma is "But the best plan is to make certain by exploration." This is very much the comment of a surgeon rather than a physician.

Investigation

He was a loyal devotee of the teachings of Hippocrates. Thus, his introductory remarks on the management of injuries repeat in principle the advice of Hippocrates to open and explore and to determine whether a fracture is present or not. He repeats Hippocrates advice on distinguishing between the sutures and a fracture, using ink as described earlier. In cases of uncertainty he advised wound exploration with probe or by incision and inspection. He insists that when an incision is made it should be X shaped for better vision and that the pericranium should be excised to avoid inflammation during healing. Hemorrhage should be controlled with a vinegar impregnated sponge. Strong vinegar was also poured onto to exposed dura to stop the bleeding.

Clinical Picture

In contrast with Hippocrates, Celsus specified the clinical consequences of cranial trauma and attempted to analyses their significance. In fact, he confused the issue by listing the posttraumatic symptoms, bilious vomiting, impaired vision, aphasia, bleeding from the nose or ears and temporary loss of consciousness and stating they only occurred if there were a fracture. He goes further to specify that if there was a persisting defect of consciousness, epilepsy or paralysis, this indicated a meningeal laceration. Thus, unlike Hippocrates, he attempted to provide a mechanism for the posttraumatic symptoms. Nonetheless, by attributing them to the wrong source his suggested notions which did not help advance understanding.

Management

Celsus specified as stated that a head injury can lead either to a fissure or a depressed fracture. In general, he was against surgical interference with fissures and advocated the use of dressings and ointments. This is much more conservative than Hippocrates. However, with a split skull the fragments could interlock preventing the escape of fluids (humors) trapped beneath, presumably produced by irritation of the dura by the fractured fragments. For this he advocated chiseling away overlapping bone margins producing a space between the fragments which would permit fluids to escape. His description of the management of depressed fractures is comprehensive, elegant, and simple and in principle no different from current practice. It is described in some detail in an earlier paper.[18]

Technique

Celsus describes how trepanation should be carried out in much the same way as Hippocrates but also emphasizes how the modiolus must be used with downward pressure the degree of which was important. If the pressure was too light the instrument would not cut. If it was too heavy the instrument would not rotate. Celsus did not however describe how this pressure was to be applied. He repeated Hippocrates advice to cool the trepan/modiolus frequently. He also mentioned that bleeding indicated that a drill had reached under the outer table of the skull. He advised about the amount of downward pressure required at different stages. He mentioned that once the groove was made the central pin of the modiolus should be removed. He specified that if the area of damaged bone to be removed is bigger than a modiolus it should be surrounded by burr holes made with a trepan and that these should be connected using a chisel so that the entire damaged area could be removed (see Fig. 3.5).

Like Hippocrates he was anxious to avoid damaging the underlying dura (cerebral membrane). He took this concern one step further by using an instrument called a meningophylax which was to be inserted

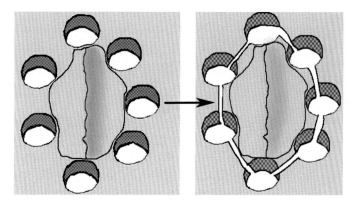

Figure 3.5 The diagram shows how multiple burr holes might be made and then joined to facilitate removal of large depressed bone fragments.

under the bone when a chisel is applied to hammer a defect between adjacent burr holes, thus protecting the dura from damage. The same plate was used to elevate the bone fragments once they had been freed. He also insisted that bone edges must be filed smooth and bone dust must be removed or else skin healing would be complicated by poor healing and new pain. The precise form of the meningophylax is uncertain but its description is elegant. He states "….. a guard of the membrane which the Greeks call meningophylax. This consists of a plate of bronze, its end slightly concave, smooth on the outer side; this is so inserted that the smooth side is next the brain, and is gradually pushed in under the part where the bone is being cut through by the chisel; and if it is knocked by the corner of the chisel it stops the chisel going further in; and so the surgeon goes on striking the chisel with the mallet more boldly and more safely, until the bone, which having been divided all round, is lifted by the same plate, and can be removed without any injury to the brain."

He mentioned that not all depressed fragments need to be removed and stated that the callus between fissures and retained depressed fractures is a better covering for the brain than the scar tissue left following fragment removal. In addition, he went into lucid detail about how to disimpact a depressed fracture from the margin of normal bone. In principle, his technique is little different from current practice and is illustrated in Fig. 3.6.

Extra Findings

Celsus was aware that blood could accumulate between the dura and the skull in the absence of a cranial fracture following a rupture of a meningeal vessel. In other words, he described an epidural hematoma. He specifies that in these cases there is local pain over the area concerned and on inspection the bone will be found to be pale and such bone should be excised.

Conclusions

It has been stated that Celsus was forgotten until he was rediscovered and De Medicina was republished in 1478. It has been said that this was because he wrote in Latin, which did not have the intellectual prestige of Greek.[19] Thus, Celsus' rational influence on surgical tradition and insights about cranial injury effectively disappeared for nearly a millennium and a half.

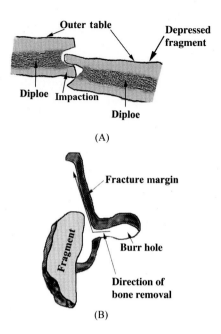

Figure 3.6 (A) A cross section through a depressed fracture demonstrating the way in which the outer table impacts into the diploe. This is a frequent feature of depressed fractures and makes their elevation more difficult. If the fragment is securely impacted it is necessary to drill an adjacent burr hole. (B) This shows a diagram from above in which a burr hole has been drilled. Then, the edge of the bone adjacent to the depressed fragment is bitten up thus freeing the depressed fragment without using unnecessary force. The author thanks the Journal of the History of Neuroscience for permission to use this figure.[18]

He admitted a debt to Hippocrates. He confirmed a trepanation technique similar to that of the master. However, he added nothing about anatomy and physiology and mentions the brain only to say he preferred callus to scar tissue to cover it after treatment. He made however serious contributions to the surgery of the injured cranium.

1. He proposed excision of pericranium to reduce the chance of inflammation.
2. He redefined fractures into fissures or depressed. He provided an elegant technique for the management of impacted depressed fractures.
3. He wrongly associated cerebral symptoms of vomiting, impaired vision, aphasia, and temporary loss of consciousness with injuries to the cranium or meninges.
4. He was aware that epidural blood could collect in the absence of a fracture and suggested overlying bone could be identified by its pallor and should be excised.

CLINICAL SCIENCE GALEN

Background

Claudius Galen's (129–216) background is lucidly outlined by RJ Hankinson in the Cambridge Companion to Galen.[20] There is quite a bit more biographical material about him than there is about his predecessors. He was born into a good family in the large and prosperous city of Pergamon, in what is now western Turkey. See Fig. 3.1. This city was reputed to possess the best library outside Alexandria. His father Nicon was an architect which would have meant also an engineer. Galen was close to his father but not his mother who had the reputation of a filthy temper, even on occasion biting her servants. This may be a reason why Galen never married and while he could relate to women he does not seem for the most part to have valued them highly. Galen would develop into an aggressive and arrogant personality and who can say how much that those tendencies were inherited from his mother?

Nicon stimulated Galen to study grammar, mathematics, logic, and philosophy. The rumor goes that he was aiming to be a philosopher when his father had a dream which decided he should study medicine. He started studies under Satyrus in Pergamon and after in Smyrna with Pelops. This was followed by a period in Alexandria. After his extensive education, he returned to Pergamon and was appointed as a surgeon to the gladiators. He was so successful at the job that he was re-elected to it by the priest in charge a total of four times. This would have given him wide orthopedic experience and would have extended his knowledge of superficial anatomy. He traveled to Rome where he worked from AD 162 to 166. He then returned to Pergamon for unclear reasons and was subsequently summoned back to Rome by the Emperor Marcus Aurelius.

He wrote an enormous amount or rather he dictated to relays of slaves, writings which cover the whole of medicine. For our present purpose, attention shall be limited to texts concerned only with operations in the head and cranial anatomy. These topics will be considered with respect to the clinic and the laboratory.

Classification of Fractures

Galen wrote the following about Hippocrates. "Of course, a whole book has been written about fractures in the head by Hippocrates .... For the present, since I add in this treatise the things discovered other than those mentioned by that man, let me define those things he stated vaguely." He reclassified the fractures as extending to the diploe, extending to the internal surface, simple, comminuted, depressed, or elevated. This is much more modern than the classification of Hippocrates and more extensive than that of Celsus.

Etiology
Galen does not discuss the etiology of cranial injury, letting it seem that he accepts Hippocrates writings on this topic.

Investigation
Again, he accepts the methods of Hippocrates and provides no detail himself.

Clinical Picture
There is no mention of the clinical picture of skull fracture. Nor is there any mention of symptoms affecting consciousness, nor epilepsy nor apoplexy. He does examine brain function in depth elsewhere but not in relation to cranial trauma.

Management
He advocated poultices to keep the wound dry in keeping with Hippocrates teachings. He was more willing to operate on depressed fractures than either Hippocrates or Celsus. Modern practice regarding depressed fractures would tend in the direction of Hippocrates and Celsus not least because Galen's method would result in extensive skull defects needing to be filled. It is noteworthy that this would not have been an option in Galen's time. The procedure of cranioplasty required to replace extensive skull defects remains a considerable challenge even today.

Galen also mentioned the dura and describes methods whereby its damage by a surgeon may best be avoided. He makes no mention of hematomas related to the dura.

Surgical Technique
He specified that raspatories should be used and introduced a new instrument called a cycliscus. This is defined in Chambers Cyclopaedia of Arts and Science, Volume 1 in 1728 as "An instrument in the form of an half-moon; used by surgeons to scrape away rottenness." The raspatories

would scrape away bone and show the depth of the fracture. He very sensibly mentioned that a large number of raspatories would be needed, indicating that he started with a broad one and then uses increasingly narrow ones as the operation moved deeper. If the fracture was not full thickness, then further surgery was unnecessary. Here again partial thickness fissures are mentioned. If the split extended through to the dura and was no more than a fissure, no more surgery would be required. If there was crushed bone it must be removed. This could be done as described by Celsus, using the trephine or using the cyclisci to scrape channels in the bone. Galen liked neither method being concerned that the trephine may penetrate too far leading to damage of underlying tissues. In spite of this, he did describe trepans fitted with a collar which would prevent them sinking in. However, this required using several such instruments with the collar placed at varying distances from the cutting end to adapt for skulls of varying thickness. He also disliked the cyclisci, in this context, because they shook the head too much. His preferred method was to use the cyclisci to get enough space and thereafter to use the lentiform knife to cut away the bone (see Fig. 3.7).

He mentions thereafter bone forceps for the removal or turning back of depressed bone fragments, but is unclear about how extensive such a removal should be. This elevation of fragments facilitated the introduction of the lentiform knife. In addition, he instructed that everything severely shattered should be removed but that it was not necessary to follow every fracture to the entirety of its extent. He introduced more instruments than his predecessors and his recommendations about their use were sensible.

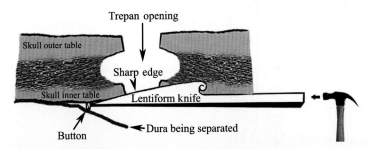

Figure 3.7 This shows the use of the lentiform knife in cross section. Space is made with a trepan to insert it under the inner surface of the bone. The dura which Galen did not think was attached to the skull is separated further to introduce the tip of the knife which has a button to protect the dura. The knife is then tapped in deeper and deeper with a mallet which results in cutting the bone while avoiding penetration of the dura with damage to the underlying brain.

Cranial, Meningeal, and Vascular Anatomy

Galen described the cranium and the meninges. Unlike his predecessors, he dissected the contents of the cranium although not in humans. He described the skull as a helmet to protect the "encephalon." The bones are joined together with interdigitating fibrous joints. He contended that the dura could not attach to the inside the skull "because their substances were so unlike."[21] Thus, the dura was attached to the sutures which in turn are attached to each other on the exterior by a common membrane, the pericranium and this attachment holds the dura in place. A clinical corollary of this notion is found in the statement "... the thick membrane very quickly separates from the bones that are severely affected."[22] This error persisted until the 18th century. This is not without significance, since if it were true, the pathophysiology of epidural bleeding would be rather different from that which has been demonstrated scientifically. Galen also described the pia mater accurately calling it the choroid membrane. He taught that it surrounded the brain and was united with it but was separate from the dura. It is insinuated into the depths of the brain and accompanies the brain blood vessels everywhere. Here there is no error. On the other hand, he went into great detail about the intracranial blood vessels. The presence of the rete mirabilis is mentioned later lying between the bone and the dura at the base of the skull. Galen considers there to be openings in the dura whence veins pass to the diploë and thence to the pericranium. Galen thought that the veins should nourish the tissue being unaware of circulation. This incorrect anatomical arrangement also persisted up to the 18th century.

Galen and the Brain Function[21]

Galen undertook extensive studies of the brain and its function. He was convinced it was the seat of the soul and that pressure to different parts of the brain could induce loss of consciousness, and this was more severe if the pressure was applied posteriorly. The same was true of incisions into the ventricles. This effect was due to influence on the soul mediated via the animal pneuma. He had nothing to say about the precise location of the soul and its nature (see Fig. 3.8).

Galen also discusses the complexity of the cerebral convolutions noting that a donkey also has extensive cerebral convolutions. From this observation, he concludes that cerebral convolutions are not related to intelligence. Rather intelligence is related to the quality of

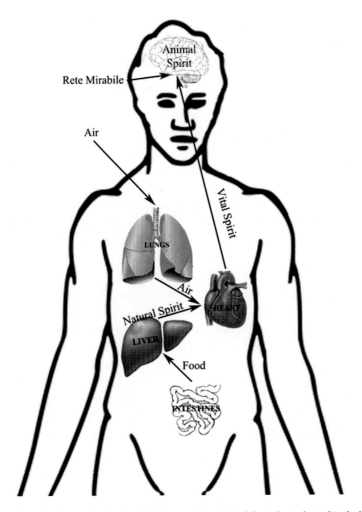

Figure 3.8 Galen believed in a three-part soul. The lusts and appetites of the soul were located in the liver. The emotionally driven virtues and moral values are placed in the spiritual soul in the heart. The rational soul which was in charge of everything else is located in the brain (see text).

the psychic pneuma rather than its quantity. He considered that intellectual functions could be divided in a modular fashion into imagination, reason, and memory. He believed these were located in the brain substance but did not specify where.[23]

His experimental findings had clinical relevance in that he had a patient from Smyrna with an injury that extended into the ventricles and the young man survived. This is comparable to observations of animals which survived incisions into the ventricle. Nonetheless, this knowledge otherwise had no systematic effect on his management of cranial injury.

Galen differed from Hippocrates and Celsus in one important respect. They observed and recorded. He experimented and analyzed and his writings are in consequence more speculative. There is one other anatomical finding of Galen's which is relevant for what follows. He described, in keeping with Herophilus seven cranial nerves. The vagus in particular, he dissected in great detail describing the recurrent nerves to the larynx and demonstrating their function in relation to the production of the voice. He also traced branches to the gastrointestinal tract; a finding which will have relevance in relation to head injuries as it would be considered the pathway for gastrointestinal disturbances, particularly vomiting which follow cranial injury.

His physiology based on humors and pneumata could not be other than speculative. Nonetheless, arising out of this physiology was one therapy which he favored and which would have immense influence on clinicians for centuries to come. This was phlebotomy. This procedure was an obsession of the medical profession for over 2000 years. Its basis was the belief that by removing blood from a patient a state of eucrasia could be re-established. Purgation would achieve something of the same benefit but was obviously less dramatic than bloodletting. It is suggested the Egyptians may have undertaken it. However, it was definitely flourishing at the time of Hippocrates.[24] Celsus stated "To let blood by incising a vein is no novelty; what is novel is that there is scarcely any malady in which blood may not be let." Epilepsy was one of the conditions were letting blood in any manner was not recommended by Celsus. Galen was very particular as to the need to perform the procedure for what he perceived as the right indications. It was necessary to remove enough blood to make the patient faint. It was however to be avoided in those whose constitution was too weak to cope with it. It is possible that the feeling of euphoria associated with blood loss, as experienced by blood donors may have contributed to perpetuating the practice.

Conclusion

Galen's contributions were legion. Regarding cranial injury, he described a safe technique and classified fractures better than hitherto.

1. Nothing specific is said about the scalp except to follow the teachings of Hippocrates about keeping cranial injuries dry.
2. He improved the classification of cranial fractures. He was more enthusiastic than Hippocrates and Celsus about removing bone fragments in depressed fractures.

3. He considered it necessary to avoid damage to the dura and described a technique by which this could best be accomplished. He also incorrectly taught that the dura was attached to the skull only at the sutures.

4. He experimented widely on the brain and cranial nerves. He described the functions and pathways of the vagus nerve and related this knowledge to clinical experience. He also showed how reversible loss of consciousness could be achieved by pressure on different parts of the cerebral ventricles. He explained this by means of a substance called psychic pneuma which communicated between the body and the soul. He did NOT relate these changes in consciousness to the clinical course following cranial trauma.

5. He made no direct contribution to knowledge about epidural bleeding.

6. As he dissected only animals he made errors in his description of the blood supply to the brain including a rete mirabilis which does not exist in primates.

Galen's work covered the whole field of medicine. His influence on succeeding centuries partly reflected his stature but also reflected his determinism which believed every bodily function was there by design, a belief that fitted well with the teachings of the Christian church.[25] In contrast with Hippocrates, he believed that wounded brain could heal as is memorized in Galen's Greek text in the ceiling of the Entrance Hall to the Montreal Neurological Institute "ἐγκέφαλον δὲτρωθέντα πολλάκις εἴδομεν ἰαθέντα."[26] This could translate as "It is known that the wounded brain can often heal".[a] It comes from his commentary on Hippocrates aphorisms.[27]

APOPLEXY HEADACHE AND EPILEPSY

Introduction

While trauma is the main indication for cranial surgery in the ancient world, there were three other conditions which could concern physicians who operated on the head. They are headache, epilepsy, and apoplexy and they should be briefly considered.

---

[a]The Author would like to thank Michael Torrens for help in clarifying the meaning of the Greek text.

Apoplexy

Hippocrates

With regard to spontaneous apoplexy Hippocrates indicates in Aphorisms Section VI No. 57, that apoplexy is commoner in people between 40 and 60.[28] He also noted Aphorisms Section II No. 42 that "It is impossible to remove a strong attack of apoplexy, and not easy to remove a weak attack."[28] In a recent review the clinical nature of apoplexy as perceived by the Greeks was "The term Apoplexia was employed by the Greeks, and is still used, to denote a disease in which the patient falls to the ground, often suddenly, and lies without sense or voluntary motion. Persons instantaneously thus affected, as if struck by lightning."[29] That is the clinical phenomenon. The pathophysiological basis of this clinical event is something else and not particularly helpful. Hippocrates did observe hemiplegias and associated a right hemiparesis with aphasia following a minor trauma.[30]

Celsus

Celsus was familiar with apoplexy and had his own take on it. He discusses it in Chapter 3 sections 26 and 27 of De Medicina.[17] His description is pithy. "...some who have been stunned, in whom the body and mind are stupefied. This is produced sometimes by lightning stroke, sometimes by disease; the Greeks call this latter apoplexia." He points out like Hippocrates that severe apoplexy is usually fatal but survivors have a poor time of it in what we would call a vegetative state. For partial paralysis, he suggests improvement is possible. He advises the usual bleeding and purgation together with avoidance of cold. More interestingly he recommends physiotherapy to the muscles, preferably active but otherwise passive. He also recommends stimulation of the skin with nettles or mustard plasters applied until the skin becomes red and then removed. This regime does seem sensible and remarkably modern.

Galen

Galen's thoughts on apoplexy were extensively reviewed in the 1990s.[31] This thorough analysis points out that information on Galen's reaction to apoplexy was scattered throughout his work and was by no means consistent. He thought it was due to an excess of phlegm. Apoplexy was the proper diagnosis in all cases of a sudden, simultaneous, complete, and nonfebrile loss of motion and sensation, including trouble of consciousness and respiratory failure. It was often succeeded by paralyses. Did Galen know about hemiplegias? This is not clear. In 1991 Charles Rose claims that he knew the term although there is no citation.[32] On the other hand, John Cooke writing in 1824 states "The disease hemiplegia, or semisideratio as some call it, has been described or adverted to by almost all the ancients, yet the word hemiplegia does not anywhere, I believe, occur in the writings of Hippocrates, Galen, or Aretaeus. Paulus Ægineta seems to have been the first who thus designated this species of palsy."[33]

Epilepsy
Hippocrates
On the matter of epilepsy Hippocrates remarks are definitive in the text "On the Sacred Disease."[9] He is definite that epilepsy is due to a disorder of the brain. However, the mechanism of that disorder is explained in terms which are no longer relevant. While Hippocrates notes that epilepsy is contralateral to trauma in his monograph on cranial trauma, there is no comment on lateralization in the text of "On the Sacred Disease." However, there is one remarkably astute and accurate clinical note about the febrile convulsions of children to be found in the Prognostics (No. 24). It is stated "Convulsions occur to children if acute fever be present, and the belly be constipated, if they cannot sleep, are agitated, and moan, and change color, and become green, livid, or ruddy. These complaints occur most readily to children which are very young up to their seventh year."[28]

Celsus
Celsus mentions epilepsy with the translation using the archaism "comitialis." He describes how patients may or may not convulse and will mostly recover. Thereafter, bleeding is to be avoided but purgation is allowed and mild foods and avoidance of stress are advised.

Galen
Galen considered that all parts of the brain were involved with epilepsy which was a disturbance located in the ventricles. This could also be due to an excess of phlegm or black bile interfering with the function of the psychic pneuma.[34] Like apoplexy there was loss of consciousness. There was also amnesia for the ictus but respiration was seldom affected and the patient soon recovered.

Headache
Hippocrates
There is almost nothing in Hippocrates writings about headache except passing mentions of its association with fever and drinking. There is no attempt at classification of different sorts of headache.[35]

Celsus
Celsus mentions a specific kind of severe headache which he calls cephalaia. He describes the condition as follows "In the head, then, there is at times an acute and dangerous disease which the Greeks call cephalaia; the signs of which are hot shivering, paralysis of sinews, blurred vision, alienation of the mind, vomiting, so that the voice is suppressed, or bleeding from the nose, so that the body becomes cold, vitality fails. In addition, there is intolerable pain, especially in the region of the temples and the back of the head .... And all these pains occur, sometimes with fever, sometimes without fever; sometimes they affect the whole head, sometimes a part only; at times so as to cause excruciating pain also in the adjacent part of the face." The remedies are nonspecific including various applications to the head and phlebotomy about which further mention is made at the end of this chapter. In conclusion, it may be mentioned that in desperate cases he recommended to apply mustard plasters over the site of the pain and if that does not work the brutal alternative of producing an ulcer with cautery may be attempted.

Galen
Galen used the terms cephalalgia for sudden head pain attributed to a temporary cause (even if there was pain for several days). Long-term chronic pain was termed cephalea. He also coined the term hemicranias and these three terms were do dominate the clinical management of headache for centuries to come.[35] He was convinced some headache had its origin inside the head in the membrane, or the substance of the brain or veins, arteries, and nerves. External pain came from other locations.

Aretaeus the Cappadocian
Aretaeus has not been mentioned hitherto because he wrote nothing about cranial injury. However, he had important things to say about neurological illnesses. He lived about the same time as Galen in the region of central Turkey known as Cappadocia. He wrote in Greek

and much of his work has survived. It is thus fascinating that he understood the crossed lateralization of cerebral functions and yet his work was forgotten or ignored for centuries.

Apoplexy

Aretaeus has a clear and detailed awareness of the clinical picture. First he stated "But apoplexy is a paralysis of the whole body, of sensation, of understanding and of motion."[36] This is in agreement with the older authors. However, he is more specific about the varieties of apoplexy. He states. "... the parts are sometimes paralyzed singly, as one eye-brow, or a finger, or still larger, a hand, or a leg; and sometimes more together; and sometimes the right or the left." It is in this section of his work that he wrote: "If, therefore, the commencement of the affection be below the head, such as the membrane of the spinal marrow, the parts which are homonymous and connected with it are paralyzed: the right on the right side and the left on the left side. But if the head be primarily affected on the right side, the left side of the body will be paralyzed: and the right if on the left side. The cause of this is the interchange in the origins of the nerves; for they do not pass along the same side, the right on the right side, until their terminations; but each of them passes over to the other side from that of its origin, decussating each other in the form of the letter X."

He has further detail of the nature of the paralyses. "The varieties of paralysis are these: sometimes the limbs lose their faculties while in a state of extension, nor can they be brought back into the state of flexion, when they appear very much lengthened; and sometimes they are flexed and cannot be extended; or if forcibly extended, like a piece of wood on a rule, they become shorter than natural. The pupil of the eye is subject to both these varieties, for sometimes it is much expanded in magnitude, when we call it Platycoria; but the pupil is also contracted to a small size."

Epilepsy

Of epilepsy Aretaeus has this remarkably humane account. "EPILEPSY is an illness of various shapes and horrible; in the paroxysms, brutish, very acute, and deadly; for, at times, one paroxysm has proved fatal. Or if from habit the patient can endure it, he lives, indeed, enduring shame, ignominy, and sorrow: and the disease does not readily pass off, but fixes its abode during the better periods and in the lovely season of life.

It dwells with boys and young men; and, by good fortune, it is sometimes driven out in another more advanced period of life, when it takes its departure along with the beauty of youth; and then, having rendered them deformed, it destroys certain youths from envy, as it were, of their beauty, either by loss of the faculties of a hand, or by the distortion of the countenance, or by the deprivation of some one sense. But if the mischief lurk there until it strike root, it will not yield either to the physician or the changes of age, so as to take its departure, but lives with the patient until death. And sometimes the disease is rendered painful by its convulsions and distortions of the limbs and of the face; and sometimes it turns the mind distracted. The sight of a paroxysm is disagreeable, and its departure disgusting with spontaneous evacuations of the urine and of the bowels." This account is far more aware of the patients' suffering than the accounts of predecessors.[36]

Headache

It is claimed in a recent book on the history of headache that the first known account of the recognition of different varieties of headache appears in the surviving writings of Aretaeus the Cappadocian.[35] It is thought by this author that the writings concerned preceded those of Galen. The first type of headache considered is called cephalalgia. "If the head be suddenly seized with pain from a temporary cause, even if it should endure for several days, the disease is called Cephalalgia. But if the disease be protracted for a long time, and with long and frequent periods, or if greater and more untractable symptoms supervene, we call it Cephalea."[37]

CONCLUSIONS

Cranial Injury

Hippocrates, Celsus, and Galen were in broad agreement in principle about head injury management, varying only in detail and emphasis. Their approaches can be summarized as follows.

1. All three surgeons had useful anatomical knowledge of the cranium.
2. Skin and integuments should be kept dry.
3. Diagnosis included a history to ascertain the probable type of injury and inspection of the wound with possibly the use of black dye to facilitate fracture detection.

4. Fractures were classified in various ways but they should be scraped to assess their depth, partial thickness fracture being an accepted lesion. Depressed fractures could be elevated but there was a difference of opinion as to how aggressive the surgery should be.
5. Useful techniques of trepanation and depressed fracture elevation are described.
6. Hippocrates and Celsus probably were aware of epidural bleeding and thought it should be removed.
7. Celsus attributed posttraumatic symptoms to injury of the bone and meninges. When his work was rediscovered in the 15th century this would be a persisting source of error for many years.
8. Galen taught the dura was held to the skull by the sutures but was not otherwise attached. This error would persist to the 18th century.
9. Humoral physiology promulgated by Hippocrates, Celsus but especially Galen was the basis for venesection and bloodletting and purgation which would remain important treatment methods into the 19th century.

Cerebral Injury

Celsus does not dwell on this topic. He attributed cerebral symptoms to cranial and meningeal injury. Hippocrates and Galen both had curiously irrelevant insights.

1. Hippocrates insisted that the brain was the seat of the intellect and emotions. He did not however equate that insight with the clinical changes following cranial injury. He was aware of the dura and the pia mater. He thought damage to the pia was not consistent with survival.
2. Galen did extensive anatomical and physiological studies of the brain. A deep-seated vein bears his name to this day. He knew that pressure to different parts of the ventricles could lead to loss of consciousness. This he interpreted as an effect on the "pneuma" which affected the soul which lived in the brain somewhere: although he expressed ignorance of where and how. However, this knowledge did not help in head injury management so that like the other ancients his practical surgery was limited to management of the bony injury. Unlike Hippocrates he insisted that it was possible to survive an injury to the pia.

Because of his authority and the respect of the early church, Galen's thought, as outlined earlier would dominate medical management of all sorts until the Renaissance.

REFERENCES

1. Lloyd GER. *Methods and Problems in Greek Science*. Cambridge: Cambridge University Press; 1991.

2. Huffman C. Alcmaeon 2013. Available from: http://plato.stanford.edu/archives/sum2013/entries/alcmaeon/.

3. Apatow R. The Tetraktys. The Magazine of Myth and Tradition. 1999; 24(3).

4. Huffman C. Pythagoras 2009. Available from: http://plato.stanford.edu/entries/pythagoras.

5. Martin L. The Story of Mathematics: Greek Mathematics Pythagoras 2010 [cited January 10, 2017]. Available from: http://www.storyofmathematics.com/greek_pythagoras.html.

6. Scarborough S. More than a theorem: Pythagoras and his brotherhood. *J West Myst Trad.* 2004;7:1.

7. Parry R. Empedocles. 2005. Available from: http://plato.stanford.edu/entries/empedocles.

8. Lorenz H. Heraclitus 2009. Available from: http://plato.stanford.edu/entries/ancient-soul.

9. Adams F. On the Sacred Disease by Hippocrates. Updated July 30, 2014. Available from: http://classics.mit.edu/Hippocrates/sacred.html.

10. Hippocrates. *Volume VIII. Hippocrates*. Cambridge, MA: The Loeb Classical Library, Harvard University Press; 1995:23.

11. Pott P. *Observations on the Nature and Consequences of those Injuries to Which the Head is Liable from External Violence*. London: Lawes L, Clarke W, and Collins R; 1768.

12. du Boulay GH. *Principles of X-Ray Diagnosis of the Skull*. London: Butterworths; 1980.

13. O'Halloran S. *A New Treatise on the Different Disorders Arising from External Injuries of the Head*. Dublin: Z. Jackson; 1793.

14. Ganz JC. Hippocrates, Celsus and Galen: head injury, the brain, and the bone. *Istoriâ Mediciny*. 2015;2(1):92−103.

15. Hippocrates. Wounds of the head. In: Wilkins RH, ed. *Neurosurgical Classics*. USA: AANS Publications; 1992.

16. Hippocrates. *On Wounds in the Head*. Cambridge, MA: The Loeb Classical Library, Harvard University Press; 1928:1−52.

17. Celsus. *De Medicina*. Cambridge, MA: Loeb Library, Harvard University Press; 1938.

18. Ganz J, Arndt J. A history of depressed skull fractures from ancient times to 1800. *J Hist Neurosci*. 2014;(3):233−251.

19. Guthrie D. *A History of Medicine*. London: Thomas Nelson and Sons Ltd; 1960.

20. Hankinson RJ. The man and his work. In: Hankinson R, ed. *The Cambridge Companion to Galen*. Cambridge: Cambridge University Press; 2008:1−33.

21. Galen. *De Usu Parium*. Ithaca, New York: Cornell University Press; 1968. 802 p.

22. Galen. *Method of Medicine*. Cambridge, MA, Loeb Library: Harvard University Press; 2011.

23. Clarke E, Dewhurst K. *An Illustrated History of Brain Function*. Oxford: Sandford Publications; 1972.

24. Parapia LA. History of bloodletting by phlebotomy. *Br J Haematol.* 2008;143:490−495.

25. Singer C, Underwood EA. *A Short History of Medicine.* 2nd ed Oxford: Oxford University Press; 1963.

26. Penfield W. *The Significance of the Montreal Neurological Institute.* London: Oxford University Press; 1936.

27. Galen. On Hippocrates' Aphorisms VI-VII, 1829.

28. Adams F. *The Genuine Works of Hippocrates.* London: The Sydenham Society; 1849.

29. Pound P, Bury M, Ebrahim S. From apoplexy to stroke. *Age Ageing.* 1997;26:331−337.

30. Hippocrates. *Epidemics.* Cambridge, MA: Harvard University Press; 1994.

31. Karenberg A. Reconstructing a doctrine: Galen on apoplexy. *J Hist Neurosci.* 1994;3:85−101.

32. Rose FC. Cerebral localization in antiquity. *J Hist Neurosci.* 2009;18:239−247.

33. Cooke J. *Treatise on Nervous Diseases.* Boston: Wells and Lilly; 1824.

34. Gross RA. A brief history of epilepsy and its therapy in the western hemisphere. *Epilep Res.* 1992;12:65−74.

35. Eadie MJ. *Headache Through the Centuries.* Oxford: Oxford University Press; 2012.

36. Aretaeus. *On the Causes and Symptoms of Chronic Disease. The Extant Works of Aretaeus the Cappadocian.* London: Sydenham Society; 1856.

37. Adams F. *The Extant Works of Areteus the Cappadocian.* Boston: Boston Milford House Inc; 1972/1856.

CHAPTER 4

From Ancient Times to the 17th Century

INTRODUCTION

After Galen, there followed centuries, the Middle Ages, where medicine was largely a matter of following his teachings, especially with regard to his humoral physiology which remained little changed. Galen believed in determinism where creatures are designed to fulfill specific functions.[1] This is consistent with Christian teaching about creation and the perfection of God. Thus, Galen was popular with the church which was the dominant intellectual authority from his time up to the Renaissance. The church believed in the superior importance of God and the spirit, to which physical sickness and its treatment was inferior. This was an authoritarian approach, with God knowing all the answers and it was a view which did not encourage either curiosity or research. Nonetheless, there were a few advances which may be mentioned. They fall into three categories; understanding of the brain, information about intracranial hematomas, and surgical technique. This chapter outlines the changes which took place over the more than 1000 years involved. They are mainly concerned with small changes in anatomy and clinical management.

Increasing anatomical knowledge in the earlier part of the middle ages could not advance since the church had inherited the Ancient world's abhorrence of dissection so this was not available.[2] The church father, Augustine (354–430) wrote strongly against human dissection. One of his friends, the physician Vindician, is quoted as saying, "It pleased the ancient anatomists to examine the viscera of the death to learn in what way they died, but for us humanitas prohibits this."[3] Useful advances in surgery could not occur until this ban was lifted because surgery is based on a knowledge of anatomy and that requires human dissection. It was the soul that mattered, not the body. Thus, while there was much effort extended in caring for patients in institutions largely connected to monasteries, academic medicine and medical research withered. In a book written in the middle of the 19th Century

Intracranial Epidural Bleeding. DOI: https://doi.org/10.1016/B978-0-12-812159-7.00004-7

Russell makes the following trenchant comment. "In short, medicine, as an art based upon the natural and ordinary course of events, was superseded for a time by the extraordinary and preternatural power of certain men. Had this power continued in the Church, then the medical profession must have entirely disappeared; for who would have gone through the painful, uncertain, and expensive methods of treatment, then and since in vogue, if all that was required to be done was to send for a holy man to pronounce certain words, and so end the distress?" He elsewhere quotes Hippocrates remarks from the Sacred Disease "To me it appears that such affections are just as much divine as all others are, and that no one disease is either more divine or more human than another; but all are alike divine, for each has its own nature, and no one arises without a natural cause." He contrasts this with the following comment "What a sad contrast to this true and admirable exposition of the causes of disease do we find in the writings of some of the most justly venerated Fathers of the Church who lived five hundred years later than Hippocrates." "It is demons," says Origen, "which produce famine, unfruitfulness, corruptions of the air, and pestilence. They hover, concealed in clouds, in the lower atmosphere, and are attracted by the blood and incense which the heathen offer to them as gods". "All diseases of Christians," says Augustin, "are to be ascribed to these demons: chiefly do they torment fresh-baptized Christians, yea! even the guiltless new-born infants."[4] Nothing was to detract from the preeminence of the one Great Physician.

Even so, ongoing interest in brain function was not entirely absent. The early church fathers, in particular Augustin of Hippo and Nemesius moved the cerebral function of reason, imagination, and memory from the parenchyma where Galen had placed them, into the ventricles. The reasons for this are not clear. Out of this change arose what is called the Cell Doctrine. The first cell, consisting of the front of both lateral ventricles was the recipient of sensation from all over the body. In Latin this function was called Sensus Communis from which the phrase common sense derives. The second cell comprised the posterior lateral ventricle and the third ventricle and was the site of imaginativa (imagination). Cogitativa (thought) or ratio (reason) were located in the middle cell which was the third ventricle. Memorativa (memory) was in the third cell or our fourth ventricle (see Fig. 4.1). Variation in the system was not uncommon.[5]

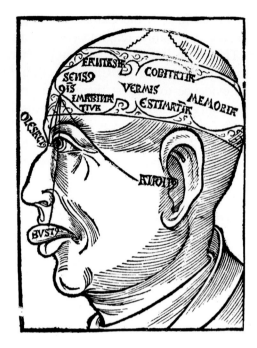

Figure 4.1 Cell Doctrine of brain function. See the text.

Fig. 4.1 also illustrates a feature of illustrations used by the early church which was inimical to good anatomy study. The Romans had mastered the technique of perspective in two-dimensional representations of three-dimensional reality as required by drawing and painting on flat surfaces. Modifications of this technique were rediscovered in the Renaissance. But at the time of the early church the stylized images had no perspective. Without perspective, accurate anatomical illustration is impossible.

GLIMMERS OF NEW KNOWLEDGE IN AND AROUND THE ARABIC WORLD

Nestorians

While the Church was hindering the advance of medical knowledge, it was suffering its own internal disputes. One Nestorius, a relatively little known priest with a talent for preaching was elected Patriarch of Constantinople. His theology was found to be heretical and he was banished. His followers fled from the city and established a church in modern day Iran. This was prior to the spread of Islam. After the

Arab conquest of Persia in 637, the Caliphate recognized the Church of the East as a millet, or separate religious community, and granted it legal protection. This is an interesting reflection where Islam showed more tolerance to Christians than the Christians did. This is especially true since there is evidence to suggest there were no real grounds for judging Nestorius guilty of heresy. The Nestorians labored for two centuries translating Greek medical texts into Arabic, thus providing a link between Greek and Arabic Medicine. It should be noted that most of the involved physicians spoke Arabic but were not themselves Arabs. Their nationalities are noted individually later.

Paul of Aegina (625–690)

Almost nothing is known of the personal experience of this man except he came from Aegina the location of which is shown in Fig. 4.2. He wrote a seven-volume epitome of the best medical practice much of which was culled from his predecessors, including Galen. He wrote of his work "On this account, I have collected this Epitome from the works of the ancients, and have set down little of my own, except a few things which I have seen and tried in the practice of the art. For being conversant with the most distinguished writers in the profession, and in particular with Oribasius, who, in one work, gave a select view of ever thing relating to health, (he being posterior to Galen, and one of the more

Figure 4.2 This shows the location of Aegina (arrow) not that far from Athens.

modern authors,) I have collected what was best in them, and have endeavored, if possible, not to pass by any one distemper. For the work of Oribasius, comprehending 70 books, contains indeed an exposition of the whole art, but it is not easily to be procured on account of its bulk, whilst the epitome of it, addressed to his son, Eustathius, is deficient in some diseases altogether, and gives but an imperfect description of others, sometimes the causes and diagnosis being omitted, and sometimes the proper plan of treatment being forgotten, as well as other things which have occurred to my recollection." An excellent review of surgical history considers Paul's contribution to have been significant and influential stating "Paul succeeded in his purpose. Because of the completeness of his work, the conciseness and lucidity of his descriptions, and the systematic organization of his books, large portions of his writings were incorporated into the texts of the principle Arabic authors. In surgery, particularly, he literally transcribed the entire body of Greek and Roman knowledge to Islam whence it ultimately returned to medieval and pre-Renaissance Europe."[6] In addition, the celebrated scholar Vivian Nutton makes two important comments on the significance of Paul of Aegina. In the first he remarks concerning the sixth book in Paul's Epitome, "Book 6 is by far the most informative (and most practically minded) Greek surgical text extant." In the second "... the number and variety of surgical instruments found in hoards such as those at Bingen, Vindonissa and, most recently, Rimini confirm the evidence of Celsus and Paul of Aegina that some surgeons at least had attained a level of competence and sophistication hardly reached again until the nineteenth century."

Paul of Aegina's treatise contains two sections related to neurosurgical topics. The first is hydrocephalus and the second is trauma to the cranium. The meaning of hydrocephalus is not quite the same as the term is understood today and the topic is outside the remit of this book. The other section concerns cranial fractures. Paul's classification is slightly different than those mentioned in the previous chapter. He describes a fissure, an incision, an expression, a depression, and an arched fracture as described by Galen. The fissure and depression are self-explanatory. The expression is comminuted. The arched fracture has fragments elevated in the middle and depressed at the edge. An incision involved the dislocation of some loosened fragment. Paul denied the existence of contra-coup fractures. He also mentioned an indentation which is not a division of the bone and thus not a fracture but a protrusion and bending of the skull inwardly forming a hollow.

Sometimes it is full thickness with separation of the dura and sometimes only affects the outer table.

His clinical approach was more clearly expressed than any of his predecessors. A cranial fracture is considered if the blow causing an injury has a nature or intensity to make it likely together with the symptoms of vertigo, speech loss, or prostration. A suspected fracture is examined using a probe. If this does not help, then black pigment is used to reveal the fracture as noted by Paul's predecessors. This is then scraped. If it does not pass beyond the diploe no further action would be needed. If it is full thickness it must be noted if the dura is separated; if fixed nothing more need be done. If the dura is separated trepanation is needed to avoid "unconcocted" pus; that is freely running ill smelling matter. This operation should in keeping with Hippocrates' instructions be undertaken within 7 days in summer and 14 in winter.

He recommended the use of a cruciate incision when performing a trepanation. He also explained how a stitch must be placed through each of the four small triangular flaps so formed so that they may be retracted by assistants giving better access to the surgeon as illustrated in Fig. 4.3. He then suggested in keeping with Galen the use of skull perforators of varying sizes and advised that fractured bone should not be removed in one piece but in fragments using a forceps or fingers. He specified the distance between the perforations and the depth to which a perforator may be allowed to penetrate. Small fragments and spicules must be removed prior to the application of dressings and he quoted Galen verbatim concerning the amount of bone to be removed.[7–9] Basically, the advice is bruised bone should be removed entirely but fracture fissures extending beyond the edge of the surgical approach may be safely left in place.

Paul also is more advanced than his more famous predecessors in some other matters. He remarked on the normal pulsation of the brain and how it diminished or disappeared when the brain expanded through a skull defect. As the underlying mechanism for these pulsations and their disappearance was not and could not be known, the observation was subsequently forgotten for centuries. Consistent with this observation he advised that dressings should be light; taken to mean not constricting. He also disagreed with Hippocrates who had been in favor of keeping all head wounds dry. Paul of Aegina advocated dressings soaked in a mixture of wine and oil.[8] It may be

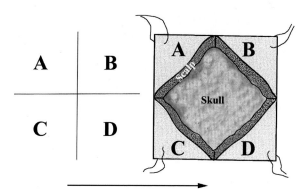

Figure 4.3 On the left is a diagram of a cruciate incision. On the right is the opening produced with such an incision. It also shows who ligatures thorough the apex of each little skin triangle may be grasped and used for retraction.

mentioned that in modern times the efficacy of both red and white wine as an antibacterial substance, unrelated to alcohol content was reported by Majno.[10] Paul lived and practiced in Alexandria. One detail of his social circumstances is worth mention. Alexandria came under Islamic governance during Paul's lifetime but there is no evidence that this had any effect upon his work.[7]

Paul also considered other illnesses related to the head including headache, epilepsy, and apoplexy. However, there is almost nothing new. He merely restated the teachings of his predecessors. No treatments were recommended that would have any chance of working. However, there was one detail. In the section titles, he used the word hemiplegia without explaining exactly what he meant by it in the text.

Abu Bakr Muhammad Ibn Zakariya Al Razi (Rhazes)

Rhazes (c. 865–932) came from Rey in the suburbs of Tehran. His major contribution to neurosurgery was the definition of concussion as a physiological state rather than evidence of permanent structural damage.[11,12] He did not however provide any explanation as a basis for this notion. The term concussion comes from the Latin concutere, which means to shake violently. An alternative name for it was commotion, also indicating its origin in shaking of the brain.[13] He also used the term hemiplegia. He also believed the brain was paired and not a single structure as Galen had believed. This is an important notion as it forms the basis for considering the expression of neurological deficit from a unilateral lesion.

Haly Abbas (c. 930–994) in keeping with Celsus and Galen advocated blocking the ears during trepanation to avoid discomfort to the
patient from the noise. He elevated a bone flap exactly as Celsus had
using multiple burr holes and a chisel. In addition, he made one important contribution to neurosurgical learning which again would be
ignored and forgotten. He proposed that intraoperative brain swelling
which could indicate the presence of further bone fragments which
should be sought and removed.[14]

Abū ʿAlī al-usayn ibn ʿAbd Allāh ibn Al-Hasan ibn Ali ibn Sīnā (Avicenna)

Avicenna (980–1037) was like Rhazes Iranian. He was a potent force
for disseminating Greek medical science into the Arabic world. His
"Canon" was very influential for many years after his death. However,
he did not make any original contribution to neurosurgical knowledge
or procedures other than translating the works of Galen.

Abū al-Qāsim Khalaf ibn al-'Abbās az-Zahrāwī (Albucasis)

Albucasis (936–1013) while Arabic was actually born in Cordova in
Spain where he spent most of his life. Much of what he wrote emphasizes the need for great skill with which he of course believed himself
to be endowed. He referred to a trephine with a collar round it to prevent it from sinking in. He also stated that many of these will be
needed to adapt to different skull thicknesses.[15] This is directly derived
from the writings of Galen.[16] His confident method of expression
meant that his book on surgery was in use for centuries.[17] While competent and confident Albucasis made no contribution to neurosurgical
thought and treatment. Finally, it may be mentioned that the illustrations in Albucasis "On Surgery and Instruments" illustrates a trepan
which would have been rotated between the hands just like the earlier
models mentioned by Hippocrates Celsus and others.[15]

Surgery, Anatomy Studies, School of Salerno and Roger Frugard

In Salerno, south of Naples on the west coast of southern Italy the first
medical school in Europe was founded. The precise nature and date of
this foundation is shrouded in mystery and obscured by legend.
Nonetheless, it maintained access to Greek texts. It was in part a lay
institution but kept on the right side of the very powerful local

monastery at Monte Cassino, a 100 miles away to the north. Later, because of Arab settlements in neighboring Sicily and the adjacent Italian coast, it is considered that Salerno was also important in the introduction of Arabic medicine into Western Europe. One of the earliest contributions from Salerno was the translation of Arabic texts into Latin making them available to physicians in Europe. The best known of these scholars was Constantine the African (1020–87) from Carthage.[18,19] The works translated were Arabic, Greek, and Arabic versions of earlier Greek texts. These translations were important bringing the knowledge back to Europe.

The scholar known as Roger of Salerno was really Roger Frugard of Parma (1140–95). He was the son of an old Parma family and wrote an erudite text on surgery.[20] This text published in 1170 and called "Practica Chirurgiae" was not written by Roger himself but by his pupils, possibly under his supervision. It became the major surgical text for at least two centuries to come.[20] There is little new in the management except one important rather clever detail. He wrote "In the case of an undisplaced fracture where one fragment does not ride below the other, one may not be able to tell if the fracture goes all the way through to the brain. To demonstrate it have the patient pinch his nostrils and close his mouth while he exhales strongly. If something appears through the crack you will know that the fracture is full-thickness."[21] If this finding was positive, it was necessary, according to Roger to make a series of trephine holes on one side of the fracture. These were then connected with a saw and resulting opening could be used to release any material trapped within.

Frederick II of Sicily was a German who also ruled parts of southern Italy. He was born in Italy of a Schwabian noble family and was Holy Roman Emperor from 1220 to 1250. In 1231 he ordained that a doctor could be licensed only after 5 years of medical study that had included surgery and had been validated by the teaching masters in Salerno. This elevation of the role of surgery was important.[19] Frederick II also is said to have proclaimed in 1238 that every 5 years a public dissection should be carried out. In 1240 he issued an edict that surgery should include 1 year of anatomy studies with a strict exam at the end.[2] This was the first formal attempt at a quality training in surgery.

Anatomy Studies and the Rise of Bologna

The practice of dissection, as mentioned earlier had been banned throughout the period of Roman domination. Thereafter, it was banned by the church.[3] The earliest mediaeval case is that caused by a Norwegian king, one Sigmurd I Magnusson (the Crusader). He arranged for one of his dead soldiers to be eviscerated in Constantinople on the way home to assess the cause of death. The appearances of the liver were similar to those of a pig marinated in the same wine the soldier had been drinking and it was considered the wine contributed to his death.[3,22] The year was 1111. The next mention was in Cremona in northern Italy. Here a monk called Fra Salimbene in 1286 during an epidemic which had affected chickens and humans. Salimbene mentions that an autopsy had been performed on a man and it was seen that he had an abscess in the same place as the hens. Emphasis is placed on the casual way this was described, suggesting that it was not an uncommon practice.[23] Nonetheless, the formal acceptance of autopsy seems to have begun with William of Saliceto, a surgeon in Bologna, who examined a man postmortem for legal reasons in around 1275.[23] Charles Springer also notes that in William's book on surgery, within the anatomy section there is a description of the contents of the thorax which could only have been acquired after a postmortem examination. Singer believed that this marks the beginning of formal anatomy study.[24,25]

Mondino de' Luzzi (c. 1270–1326), a native of and a medical student and later a Professor in Bologna had written an anatomy text book "Anothomia" published in 1316. Fig. 4.4 is taken from that book. This was not a particularly good book but it marked an important change in attitudes. It was based on Mondino's own observations no matter how inaccurate these may have been. He was the first man to have performed human dissection to study anatomy in over 1700 years. Thus, dissection was being undertaken in Bologna, no matter whether permitted or not but in 1405 the procedure received official approval in the University statutes.[2] Thus, Mondino may not have contributed much to anatomical knowledge himself but his approach to anatomical studies including the fact that he dissected himself initiated a process which would not be repeated until Vesalius but which nonetheless was an essential first step on the way.

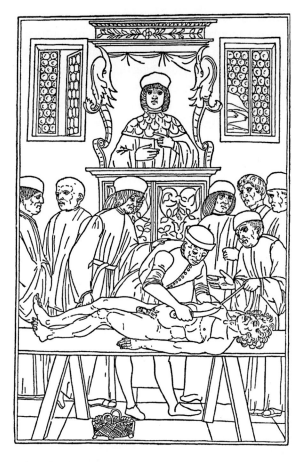

Figure 4.4 A picture of Mondino demonstrating. At this time, he had become so senior he sat in his professor's chair and commented on the dissections of others. Earlier in life he had done the dissections himself. A revolution!

Theodoric of Cervia (Borgognomi)

Theodoric (1205–98) was famous for his clean surgery and abhorrence of "laudable pus" which inevitably accompanied healing by secondary intention. He wrote an adequate account of the treatment of cranial injury in keeping with his time.[26] There is however one remarkable passage the importance of which failed to be noted until centuries later. He is describing the clinical consequences of cranial trauma. "The bad symptoms are: the loss of reason, unconsciousness, aphasia, suffusion of the face and proptosis, loss of appetite and digestion, constipation, violent diarrhea, acute fever and convulsions and the like; and these are observed in many cases which are incurable."

And very often the dura mater is torn, and sometimes the pia mater too, and the above mentioned symptoms do not appear; on the contrary, the membranes are repaired and the patient is cured. Wide experience gives us confidence in this regard, for we have seen many cases with both membranes broken, and some from which no small quantity of the brain tissue has issued, become completely cured ".... I have known a man, whose wound cavity was wholly emptied of brain substance, and finally refilled with flesh in the place of brain substance, and he was cured by Master Hugo. And since it had been the cavity of memory, I saw Master Hugo was greatly amazed over this, for the man had memory just as before; for he was a chair maker, and lost only his skillfulness." The memory refers of course to the Cell Doctrine mentioned at the beginning of this chapter.

The finding that someone can survive loss of brain substance is striking. Hippocrates in Places in Man had stated that "Of the brain there are two membranes, the fine one cannot be healed, when once injured." Galen had on the other hand referred to a young man from Smyrna who had survived an injury which involved the ventricles; indicating brain injury was survivable.

William of Saliceto

William of Saliceto (1210–77) is a figure of whom almost nothing is known. He was born in the village of Saliceto. Nothing more is known of his biography but he practiced and taught surgery in Bologna. In 1275 his "Chirugia" was published and shortly after he died. This book contains a short section on trepanation and also a section on anatomy. The clinical features leading to a diagnosis of cranial fracture differ little from those of Paul of Aegina. The book contains a brief section on applied anatomy including that of the head. William repeated Galen's error that the dura being attached to the cranium only at the sutures and hence emphasized the need to avoid trepanation at the sutures because of the risk of perforating the dura.[25] However, what makes the book interesting is a clinical detail. "...and the blow was strong enough to strike down the victim and cause him to lose his powers right then or afterwards, the paralysis, if any, will be on the side opposite the injured side of the head."[25] This is again acknowledgment of contralateral function which clearly was observed and yet which would not become part of general teaching in the centuries to follow.

Guy de Chauliac

The last innovative mediaeval surgeon who made new contributions was Guy de Chauliac (1300–68). He has three sections in his book "Major Surgery" devoted to the cranium and its contents.[27] The first is concerned with anatomy and the second with trauma and the third with aposthems. These are defined as swellings with abnormal contents. An aposthem with pus is an abscess. de Chauliac classified them into a mystic terminology with its roots in humoral physiology. In relation to the head, there is no management that relates to the topic of this book and the terminology is quite obscure.

The anatomy of the cranium, meninges, brain, and cranial nerves does not differ in any significant respect from that of Galen. Again, there was the notion that the pericranium is engendered from the dura via the sutures. He repeated the Cell Doctrine physiology for the functions of the ventricles within the brain.

While de Chauliac's treatment of head injuries did not introduce much that is new, his writings provide instruction which is more detailed and clearer than his predecessors. He mentioned the presence of water leaking when the patient sighs or holds his breath. This is the first mention of what was probably cerebrospinal fluid since the Edwin Smith papyrus. He did not however comment on the meaning of such a finding. He quotes the various authorities in detail and informs about their varying opinions. Only one detail has relevance for epidural bleeding. He specified eight elements of usefulness in operations. The third states "Whenever possible make your incisions to avoid the commissures. That is where the dura mater is attached to the bones and it may fall away or be injured, thus specifying the consequence of the error of dural attachment persisting from the teachings of Galen."

CONCLUSIONS

This chapter covers the nearly 1200 years from the death of Galen to the legalization of dissection in Bologna. Important new or rediscovered concepts include the following. It is possible to survive loss of brain tissue. The normal brain pulsates but this stops when the brain expands through a skull defect. An injury on one side produces a loss of function on the opposite side. Performing a Valsalva maneuver can give a clue as to whether a fissure is full thickness or not. None of these findings passed into the general corpus of medical knowledge of succeeding generations.

REFERENCES

1. Hankinson RJ. Philosophy of nature. In: Hankinson RJ, ed. *The Cambridge Companion to Galen.* Cambridge: Cambridge University Press; 2008:210−241.

2. Lassek A. *Human Dissection, Its Drama and Struggle.* Oxford, England: Blackwell Scientific Publications; 1958:58−60.

3. van den Tweel JG, Taylor CR. The rise and fall of the autopsy. *Virchows Arch.* 2013;462:371−380.

4. Russell JR. *History and Heroes of the Art of Medicine.* London: John Murray; 1861.

5. Clarke E, Dewhurst K. *An Illustrated History of Brain Function.* Oxford: Sandford Publications; 1972.

6. Zimmerman L, Veith I. *Paul of Aegina (Seventh Century A.D.). Great Ideas in the History of Surgery.* New York: Dover Publications; 1967:74−75.

7. Tecusan M. Paul of Aegina. In: Bynum W, Bynum H, eds. *Dictionary of Medical Biography. 4.* London: Greenwood Press; 2007:985−986.

8. Aegina P. *On Fractures of the Bones of the Head. The Medical Works of Paulus Aegineta, The Greek Physician, Translated into English. 2.* London: J Welsh, Treuttel, Wurtz & Co; 1834:429−442.

9. Adams F. The Seven Books of Paulus Ægineta 1846.

10. Majno G. *The Iatros. The Healing Hand.* Cambridge, MA: Harvard University Press; 1965:141−206.

11. Souayah N, Greenstein J. Insights into neurologic localization by Rhazes, a medieval Islamic physician. *Neurology.* 2005;65(1):125−128.

12. McCrory P, Berkovic S. Concussion. The history of clinical and pathophysiological concepts and misconceptions. *Neurology.* 2001;57(2):2283−2289.

13. Ganz J. Head injuries in the 18th century: the management of the damaged brain. *Neurosurgery.* 2013;73(1):167−176.

14. Goodrich JT. Landmarks in the history of neurosurgery. In: Ellenbogen RG, Abdulrauf SI, Sekhar LN, eds. *Principles of Neurological Surgery.* 3rd ed. Elsevier; 2012:3−36.

15. Albucasis. *Albucasis on Surgery and Instruments—A Definitive Edition of the Arabic Text with English Translation and Commentary.* Berkeley, CA: University of California Press; 1973.

16. Galen. *Method of Medicine.* Cambridge, MA: Loeb Library: Harvard University Press; 2011.

17. Guthrie D. *A History of Medicine.* London: Thomas Nelson and Sons Ltd; 1960.

18. Zimmerman L, Veith I. *Salernitan Surgery. Great Ideas in the History of Surgery.* New York: Dover Publications; 1967:93−99.

19. Nutton V. *Medieval Western Europe 1000−1500. The Western Medical Tradition.* Cambridge: Cambridge University Press; 2009:139−205.

20. Rosenman LD. Roger Frugard (AKA Roger Frugard of Parma). In: Bynum WBH, ed. *Dictionary of Medical Biography (Volume A-B).* London: Greenwood Press; 2007:523−524.

21. Frugard R. Linear cranial fractures. The Surgery of Roger Frugard Translated from the Latin Venetian Edition of 1546: XLibris; 2002. p. 39.

22. Malmesbury W. *Gesta Regum Anglorum*. London: Sumptibus Societatis; 1840.

23. O'Malley CD. *Andreas Vesalius of Brussels 1514–1564*. Berkeley: University of California Press; 1964.

24. Singer C. *A Short History of Anatomy from the Greeks to Harvey*. New York: Dover Publications Inc; 1972.

25. Rosenman LD. *The Surgery of William of Saliceto an English Translation*. Philadelphia: Xlibris Corporation; 2002:275.

26. Campbell E, Colton J. *The Surgery of Theodoric ca. AD 1267 Book 1*. New York: Appleton-Century-Crofts Inc; 1955.

27. Rosenman LD. *The Major Surgery of Guy de Chauliac an English Translation*. Philadelphia: Xlibris Corporation; 2005:692.

CHAPTER 5

The Renaissance

INTRODUCTION

In surgery as all else, the Renaissance was a period of rebirth. For the management of head injuries, the first new ideas were to come from Italy. The major anatomical revolution would come in 1543 with the publication of the work of Vesalius. However, there is one important and largely forgotten anatomist and surgeon who preceded him.

ITALY—GIACOMO BERENGARIO DA CARPI

Giacomo Berengario Da Carpi (1460−1530) was the son of a barber−surgeon Faustino Barigazzi but he followed the custom of the time and took his surname from the village of his birth. He was born just 10 years after the invention of the printing press in 1450 and thus at the very beginning of the Renaissance. He learned surgery as a youngster from his father. Berengario da Carpi is a good example of a man whose talent was not matched by virtue. Despite his initial advantages, he seems to have been a thoroughly unpleasant individual. It was said of him that, "Avarice and cupidity stir in the mind of this great man greater passions than glory and virtue."[1] In 1511 together with a servant and 16 companions he attempted to catch a man he did not like with a view to harming him. Failing to find him he wanted to go to his parents' home shouting "To his home. Let us kill his father and mother." Failing to enter the parents' house he damaged the exterior.[2] Despite his spectacularly aggressive behavior he was a popular teacher who filled his lecture theater. He was also skilled at retaining the patronage of important people, not least members of the Medici family. This probably kept him out of trouble when his temper ran away with him. However, it is happily not his personality but his talents and contributions which concern us here. Despite not being a native such was the respect in which he was held that in 1502 he was appointed to a lectureship in medicine in the University of Bologna.

Intracranial Epidural Bleeding. DOI: https://doi.org/10.1016/B978-0-12-812159-7.00005-9

Berengario Da Carpi wrote a great deal about various aspects of cranial trauma. His book "De Fractura Cranii" (On Fractures of the Cranium), published in 1518[1] was the first purely neurosurgical text since Hippocrates wrote "On Wounds in the Head" approximately 2000 years earlier. He quotes Celsus repeatedly in this book indicating that Celsus' writings had become part of the medical curriculum. This is consistent with the republication of Celsus' De Medicina in 1478, following its rediscovery by Pope Nicholas V. Da Carpi also wrote an anatomy textbook called Isagogae Breves in 1522.

The anatomy book was based on the personal experience of a great number of dissections of humans and disagreed with earlier texts on several points. To begin with there was no rete mirabile. He never found one in a human. Particularly, important in the current context, he insisted that the dura was attached all over the interior of the skull and not just at the sutures. This meant that it was not kept suspended from the pericranium through the sutures. This is anatomically accurate and it also meant that Da Carpi would be very concerned that trepanation should only be undertaken when the dura was separated from the skull; something to which we shall return. Thus, based on the personal dissection of humans he could start the process of rewriting the teachings of Galen.

His clinical teaching was very similar to that of the ancients, believing like Celsus that meningeal injury was associated with severe persistent headache, red eyes, and vomiting. He considered injury to the dura was less serious than injury to the pia and that pial injury was almost inevitably associated with injury to the brain parenchyma. Brain injury was associated with obstruction of reason, loss of vision, tremor, and shaking. He also stated that brain injury could occur without the development of "bad" symptoms, which is consistent with the finding of Theodoric (in the previous chapter) that patients can survive loss of brain tissue. In terms of one side of the brain controlling the opposite side of the body, he mentioned that injury on one side could give paralysis on the opposite side. He seems to have been a little unclear on the matter of concussion associating it with a tear in brain nerves. On the other hand, he thought that leakage from a torn "vein" would give symptoms after some delay, which is an early mention of the notion of a *lucid interval*. Finally, he considered that a blood vessel could be injured under the skull in the absence of a fracture. He wrote

"Celsus among others holds this opinion of the veins in his chapter on the cure of the skull in these words: 'Sometimes, however, it happens that the bone at any rate remains intact within but from a blow some vein in the membrane of the brain is ruptured and emits some blood....'" He then a little later in the text states his own experience. "But someone will wonder how it is possible that when the bone is not fractured the dura mater is separated from the cranium. I say and, reader, pay close attention, that sometimes by contusion some little vein contained between it and the bone is ruptured in the dura mater from which blood flows and putrefies therein. By necessity those small ligaments of the dura mater which are attached to the bone also putrefy and thus the membrane is separated and suffers serious symptoms." Here we have ideas quite close to modern concepts. Having observed that the dura is adherent to the skull everywhere, Da Carpi provides a mechanism whereby it may separate producing an epidural space into which material can accumulate.

Da Carpi mentions the relationship between the side of the trauma and the side of paralysis. He states the following: "Note that Avicenna, Canon 1,3, says that paralysis occurs on the side of the wound and spasm on the side opposite to it as in many cases. Nevertheless, I say that the reverse situation can happen, that is, in the injured part spasm can occur while paralysis occurs in the part opposite. Likewise it is possible that only one of these afflictions may occur or neither."[1] Thus, according to Da Carpi the paralysis can be on either side.

Da Carpi also had much to say about the indications and timing of the surgical treatment of head injuries. If there was penetration of the brain, then urgent surgery was required. Otherwise he was concerned with the accumulation of putrefaction within the cranium. In the event of the accumulation of putrefied material within the head after trauma, it was expected that symptoms would arise after 1 week in the summer and 2 weeks in the winter. In the event of such deterioration, there is much discussion of where a trepanation should be located. There are illustrations of the instruments he used and he mentions that when necessary he designs new ones. These illustrations are the first ever published of specifically *neuro*surgical tools. This was particularly relevant for depressed fractures. The elevator he developed for these lesions is illustrated in Fig. 5.1. With regard to these fractures

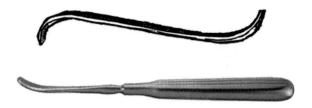

Figure 5.1 Above, *Berengario da Carpi's instrument for elevating depressed fracture fragments.* Below, *an Adson periosteal elevator which is used for the same purpose today.*

Figure 5.2 This is the diagram of a trepan from Da Carpi's book.

Da Carpi wrote "If however symptoms supervene the bone must first be perforated with some drill placed in a healthy spot in the neighborhood of the compression by the bone, making a foramen large enough to position an instrument that can elevate the bone. If it happens the entire depressed bone is separated from the dura mater and from the pericranium it should be completely removed. If instead it is firmly attached to both or to one of them it should be elevated to its original position."

Da Carpi also illustrated the instruments he used one of which is shown in Fig. 5.2. It would seem to be the first brace and bit type instrument thus replacing the hand rotated modiolus derivatives. It for the first time permitted precise control over the downward pressure exerted by the operator himself.

Da Carpi's also had advice for the treatment of injuries to the brain and the meninges. "But since the subject of the lesion of the dura and *pia* mater as well as of the medullar substance of the brain has not been treated on any method of cure by any author in writing and in order that my book may be more complete and useful it appears to me that something must be said concerning wounds of the aforesaid members especially since some times as I said before such wounds have come under my hands which with the help of divine guidance have been brought to a fortunate end."[1] He advised that such wounds should be interfered with as little as possible. Even pus should not be wiped away manually but removed by irrigating with lotion. He considered that after about a fortnight the wound might be handled more boldly albeit always gently.

Considering, the views of all the anatomists and surgeons who had preceded him Da Carpi's insights represented a major advance in understanding. However, as has happened all too often his advances would not be generally accepted for many years. Also unlike his successors, including Vesalius he did not use graphic illustrations even though others, especially in Germany were doing so.

DA CARPI'S CONTRIBUTIONS

Da Carpi made a number of contributions, the significance of which was ignored.

1. The dura was evenly attached to the whole of the inner surface of the skull and thus would be close to the bone wherever that was perforated surgically, requiring great care anywhere in the head.
2. He thought meningeal injury was associated with headache and vomiting while brain injury was associated with loss of reason, loss of vision, tremor and shaking, An insight centuries before its time.
3. He agreed with Theodoric that brain damage did not always lead to clinical deficit.
4. He thought leakage from a ruptured intracranial vessel could cause delayed clinical deterioration. Like Celsus he thought it was possible for this to happen in the absence of a fracture.
5. He considered posttraumatic paralysis could be either ipsilateral or contralateral.
6. He developed an elevator to help elevate and/or remove depressed bone fragments.

HIERONYMUS BRAUNSCHWIG AND HANS VON GERSDORFF

The German neurosurgical developments were naturally enough limited to trauma surgery and they did not make a contribution to the management of epidural bleeding. Nonetheless, a brief mention is necessary. At the turn of the 15th–16th century quality illustrations had become available following Gutenberg's invention of a printing press with moveable type. Thus, a Hieronymus Braunschwig (c. 1450–1512) who titled himself "Surgeon of the Free Imperial City of Strasburg." Could published in 1497 "Das Buch der Cirurgia" (The Book of Surgery; Braunschwig, 1497). Printed by Hans Gruninger, it was the first ever German language printed medical textbook and the first medical text to use extensive detailed woodcut illustrations.

At the end of the 15th and during the 16th another German, Hans von Gersdorff (c. 1455–1529) made a significant contribution to the management of depressed skull fractures. Little is known of his personal life other than he too came from Strasbourg and was an experienced battlefield surgeon. Nonetheless, his character would seem to have been exemplary. From his book, in a passage on the correct behavior of the surgeon he writes, "it is necessary that the surgeon have good intelligence and understanding. Not too rash in his actions, but always well aware of the harm that might come to him or to the patient because of his lack of skill." He also wrote, "Also guard yourself, if an injury comes to you which you do not understand how to heal it, you should willingly direct him away to another experienced Master."[3] Von Gersdorff published a book called Feldbuch der Wundarznei (Field Book of Wound Medicine) in 1517. His printer was Johannes Schott also of Strasbourg. The woodcuts in this book are more dramatic than in the Braunschweig book and they show surgical technique including machines that are supposed to elevate fragments in cases of depressed fractures of the cranium (von Gersdorff, 1517/1967). It may also be noted that many complex metal instruments are illustrated in the book for treating fractures and dislocations, indicating a passion for engineering applied to surgery.[4] The details of his method have been published elsewhere.[5] Among his inventions was an instrument for elevating depressed fractures called a tripes because it had 3 ft. It did not and could not work and was the subject of negative comment in later generations. Nonetheless, it would seem the rapid spread of information permitted by the printing press taken together with the very dramatic woodcuts led to it becoming a firm favorite. It was a method which did nothing to improve the management of patients with cranial trauma. It was also an early example of the power of the dramatic pictures in print.

VESALIUS

In spite of the attractive, dramatic but less than helpful German texts, the improvements in publication technology led to the production of a vital book in the mid-16th century which would form part of the basis for modern medicine and which would enable escape from the stranglehold of the church supported authoritarian learning of the ancient world. This book was the monograph "De Humani Corporis Fabrica" published in 1543 by Andreas Vesalius (1514–64).

Every physician learns that modern anatomy began with Vesalius. However, we are not routinely taught about how this came about. Vesalius was born into a distinguished medical family with contacts with royalty and universities. He started his formal education in Louvain in 1528 at the age of 14 and his major studies were the languages of Latin, Greek, and Hebrew. It is recorded that while his Latin was fluent his Greek was patchy and his Hebrew nonexistent. Nonetheless, fluent Latin gave him access to a wide range of knowledge. In 1533 at the age of 19 he moved to Paris to begin his formal medical education. Paris was prestigious but conservative and dissection, which had become legal, was only rarely practiced. Medical knowledge came from Arabic sources and from Arabic translations of the ancient masters. However, this was the time of the appearance of new translations of Galen directly into Latin so that a better quality of writing became instantly available. It is widely believed that Vesalius was opposed to Galen's teachings but this is not the case. Rather he refused to accept them slavishly as gospel and rather tested them against the results of observation and noted discrepancies. While there was little dissection in Paris in his 3 years there he increasingly took part in the presentation of dissections. In the tradition established 200 years earlier in Bologna by Mondino and continued in the same place at the beginning of the 16th century by Giacomo Berengario da Carpi, he performed his own dissections and based his writings on what he observed personally.

Vesalius and his colleagues visited the Gibbet of Montfaucon where executed bodies were brought to rot to a condition where they could be disposed of. It was haunted by crows and pariah dogs but was a rich source of material for avid anatomists. There was also the Cemetery of the Innocents whence as Vesalius remarked there was "an abundant supply when I first studied the bone." This is so far removed from the conditions of study in a modern medical school, even for those undertaking the dissection of corpses within clean carefully contained systems of storage.

He moved back to Louvain because of war and following his qualification he was appointed as professor surgery in Padua at the tender age of 23. This was an appointment which required the teaching of anatomy as well. He was popular and influential because he did not delegate dissection to others but undertook it himself. This was novel and popular and his classes were crowded. Moreover, he wrote down and published his anatomical findings with elegant illustrations. He found that his students welcomed these illustrations and in 1538 he published six large plates called the Tabulae Sex illustrating the skeleton from front, back and side together with the veins arteries and liver in a way to illustrate the physiology of Galen. Here again was evidence of the power of the printing press in disseminating information as never before.

Vesalius remained a thorough student of Galen and contributed to the publication of the old masters works in Latin in 1541. If nothing else this is evidence that Vesalius did not oppose all Galen's teachings. It is also noteworthy that he contributed to the work on Galen when his own masterpiece, De Humani Corporis Fabrica was in preparation. This effort bespeaks an unusual capacity for work. This is not the place to go into detail about the book but it is important to note that it used illustrations by major artists and that there was an elaborate system of cross references between text and illustration which at the time was revolutionary.

Vesalius' "Fabricia" of course included Section 4 on the brain and the nerves. In this section and in the "Tabulae Sex" illustration of arteries is the clearest evidence of Vesalius' debt to Galen. The arteries illustration shows the rete mirabile which does not exist in humans and the absence of which had been demonstrated by Berengario da Carpi. The illustration of the base of the brain and the classification of the cranial nerves is essentially unchanged from Galen and in the case of the cranial nerves from Herophilus. Vesalius' anatomy of the brain has been translated by Charles Singer who shows interesting insights.[6] To begin with he lets it be known that Vesalius may have been fluent in Latin but his style was obscure and repetitious, hindering understanding. He tells us that the section of the Fabrica on the brain is the best description and depiction hitherto. However, there are errors. Comment here is limited to those aspects of the text which have relevance for cranial trauma. As indicated earlier in the section on angiology the presence of a rete mirabile is still described. In respect of the dura he writes "We must be the more cautious when we handle the cautery or {treat} wounds of the head lest, perforating the skull, we disastrously penetrate {the dura} with the bone for ... at the sutures the membrane barely leaves the bone." Indicating that the dura is less closely attached elsewhere and persisting in the error previously described many times.

The illustration of the base of the brain (Fig. 5.3) does not look of the same quality as that of the rest of the Fabricia. One could speculate that this had been made by Vesalius himself but nobody knows.

At any event while Vesalius produced beautiful images of the brain he still perpetuated a number of errors which go back to Galen and beyond.

AMBROISE PARÉ

Unlike the Germans mentioned earlier Ambroise Paré (1510–90) remains a celebrated surgeon even to the present day. Paré's writings covered most of surgery including cranial injuries, though not epidural hematomas. His approach to the clinical picture following cranial trauma is much influenced by Hippocrates, Galen, and Paul of Aegina. In the history, he adds little new. In terms of the examination following Roger Frugard, he advocated using a valsalva maneuver and if blood or seropurulent fluid flows out of a fissure, then the fissure is full thickness. When a fissure is to be scraped to assess its extent, the amount of bone removed should, in keeping with Celsus' view, be as little as possible and just enough to permit the passage of blood and/or sanies. The fear is that if these are not removed putrefaction will occur, leading to a bad outcome.[7] This is based on the ancient notion that

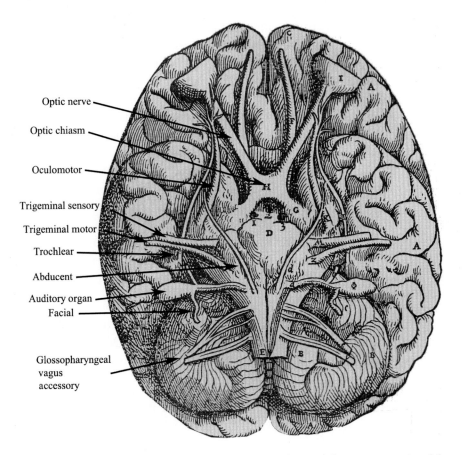

Optic nerve

Optic chiasm

Oculomotor

Trigeminal sensory

Trigeminal motor

Trochlear

Abducent

Auditory organ

Facial

Glossopharyngeal
vagus
accessory

Figure 5.3 Base of the brain as viewed by Vesalius with seven cranial nerves and a poor representation of the pons.

coagulated blood will become spontaneously putrefied if left to itself. His work became rapidly available to surgeons in France and other countries, and his reputation became most respected.[8] It must be remembered he was the surgeon to four successive French kings.

Paré's knowledge of classical writings, despite his lack of a classical education was comprehensive. Since the introduction of the printing press, publication had become a speedy process since there were many translations of the classics into French in his day.[a] Paré was the first distinguished surgeon to publish in French: behavior which was unpopular among the physicians but which probably led to the speedier spread of

[a]The author is grateful to Professor Boleslav Lichterman of I.M. Sechenov First Moscow State Medical University, Moscow, Russia for this detail.

his teachings. He had no medical degree but had qualified as a barber surgeon, who during his training had spent 3 years at the Hôtel Dieu in Paris. His impact during his lifetime was great in part because of the availability of the printing press and woodcuts to convey his teaching efficiently and speedily to his colleagues. Also, he had a fine personal reputation and wrote with simplicity and clarity.

In the current context, he believed concussion was due to shaking of the brain damaging blood vessels producing either an epidural or subdural hematoma. In this context, he mentions the pia mater. This writing is in keeping with Da Carpi and like him Paré considered epidural and subdural hematomas were the cause of symptoms. He did not seem to regard them as lesions to treat; more postmortem findings. Paré was also aware that such injuries could be the result of minor trauma as in one of Hippocrates cases quoted in Epidemics Book 5.[9] Hippocrates may have suggested letting out blood from under the bone in the temporal region and Celsus may have noted collections of epidural blood can occur without a fracture. Neither mentioned the clinical concomitant of such bleeds. Paré's work agreed with Da Carpi in believing that hematomas were a cause of loss of consciousness. In contrast with Da Carpi he advises against trepanation of the sutures because of he seems to have accepted the Galenic notion of dural attachment to the sutures and thence to the pericranium.

He opened the skull using both two-handed trepans which were cylindrical mounted on a gimlet and a one-handed trephine with a conical bit and with cutting blades on the edge as well as the tip of the instrument.

At the time of his death while the seriousness of changes in consciousness was appreciated, the origin of such symptoms remained unknown. Active surgery continued to be directed toward fractures with more sophisticated if not necessarily more efficient instruments. Brain management remained palliative at best. Paré was aware of contralateral epilepsy following a head injury but his explanation is obscure and just plain wrong. He does not comment on paralysis being contralateral.

It may be mentioned that some of Paré's success may be the result of an extraordinary shrewdness, which may be illustrated by a comment of his concerning the postoperative management of neonatal inguinal hernia. He comments "... the chief of the cure consists in folded clothes, and Trusses, and ligatures artificially made, that the restored gut may be contained in its place, for which purpose he shall keepe the child seated in his cradle for 30 or 40 days, as we mentioned before; and keepe him from crying, shouting, and coughing." If the operation was not a success, failure on the part of the patient to comply with the earlier instructions provides a splendid excuse for that failure.[10] That is very clever advice.

However, despite his skill and wisdom, Paré did not make any contribution relevant to the management of epidural bleeding. He did however contribute to persist the notion that substantial portions of the brain could be lost and the individual concerned could yet function. He quotes three relevant cases. Second, he mentions the case of one Guido of Caulia who said "he saw one which lived and recovered after a great portion of the brain fell out by reason of a wound received on the hind part of his head." Third was his own case "while I was Chirurgeon to the Marshall of Montejan at Turin, I had one of his Pages in cure, who playing at quoits received a wound with stone upon the right Bregma with a fracture, and so great an Effracture[b] of the bone, that the quantity of half a hasell Nut of the brain came forth." First, the English translation of Paré's text reads "Galen affirms that he saw a boy in Smyrna of Ionia that recovered from a great wound of the brain, but such an one as did not penetrate to any of the ventricles." Paré correctly annotates that the case in question comes from Galen's "De Usu Partium" Book 8. However, the translation of the relevant portion of Galen's text reads as follows. "At Smyrna in Ionia I once saw an incredible sight, a youth who had suffered a wound in one of the anterior ventricles and yet survived by the will of God." Thus, it would seem that Paré is misquoting Galen. However, on reviewing the original French the text is "pénétrant jusques à l'un des ventricules antérieurs." This formulation is ambivalent which is why the translation is incorrect.

PETER LOWE

A younger contemporary of Paré was Peter Lowe (1550–1610) of Scotland who was educated in Paris and ended up as surgeon to Henry

[b]Effracture is a depressed fracture.

IV of France the successor to the last of the four French kings whom Paré had served. He wrote a text entitled "The Whole Course of Chirurgerie" which was published in London in 1597. It contains a short section on cranial trauma.[11] He too was aware of contralateral epilepsy and mentions apoplexy or paralysis. He mentions the appearance and movement of the dura just like Paul Aeginata. He also mentions that if a fracture is palpated, the patient should hold their breath. If there is "issue out humidity," then the fracture extends through both tables just like Roger Frugard. He advises, like Paré and quoting Celsus that only as much bone should be removed as to permit the issue of underlying blood or matter.

Lowe describes concussion which he attributes to the same mechanism as Paré which is rupture of intracranial arteries and veins. He describes the risk of accumulating both epidural and subdural blood though he does not use those terms nor does he consider removing the collections of blood to relieve pressure. On the contrary, he attributes the risk of a blood collection as due to the possibility of it rotting and producing inflammation and not in terms of the raised pressure which concerns us today. In wounds of the head, he advises abstinence from strong drink and to eat "little and of good digestion." He then lists the constituents of such a diet which are "Capons, Chickens, Pigeons, Veal, Mutton and such like." He advises purgatives and bleeding from the cephalic vein, an advice he attributes to Paré.

He persists in the notion that fissures may be limited to the outer table of the skull, or that and the diploe but without affection of the inner table. This notion which goes all the way back to Hippocrates is of course unknown in modern practice. The question arises as to whether belief in its existence is due to faulty observation, a different kind of wound that we no longer see or to examining every possible fracture by opening the scalp and looking. The question has to remain open because even into the subsequent two centuries, belief in partial thickness skull fractures persisted. His technical advice for trepanation is very similar to that of Hippocrates.

CONCLUSIONS

The coming of the Renaissance brought printing and increasingly independent minds. Yet reviewing this chapter shows yet again how it is not enough just to improve knowledge. There is the question of persuading colleagues to accept the findings. Thus, Berengario Di Capri had understanding far ahead of his time in respect of cerebral lateralization, the relationship between the severity of symptoms and the depth of an intracranial hematoma and the attachment of the dura to the skull. Surgeons continued to note that leakage of blood or fluid from a cranial wound during a Valsalva maneuver, as first noted by Roger Frugard indicated a full thickness fracture, without really understanding the underlying mechanism.

German surgeons demonstrated greatly improved publication and textbook illustration and the results of their publications underlines how the dramatic appearance of illustrations can have more influence than the reasoned observations of such as Da Carpi. The German mechanical methods for the treatment of depressed fracture were consistently useless. Paré also considered that hematomas between dura and skull and dura and pia mater could give headache and vomiting. The latter he attributes to the connection between the brain and the stomach via the vagus nerves. He noted that epilepsy could be contralateral to an injury but did not mention paralysis. His explanations for contralateral epilepsy were not correct. Paré also noted that if a valsalva maneuver produced a flow of blood or infected fluid through a fracture, then that fracture would be full thickness. (He like all his predecessors believed in partial thickness fractures). Peter Lowe's description of the use of the valsalva maneuver in this context is consistent with observing the flow of blood and cerebrospinal fluid. He also seems to be the first to have noted the movement of the dura at surgery since Paul Aeginata.

All the surgeons who were concerned about the accumulation of blood within the cranium were worried not by the pressure but by the possibility that the blood would degenerate and become purulent with disastrous consequence.

REFERENCES

1. Lind L. *Berengario da Carpi on Fracture of the Skull or Cranium (Translation)*. Philadelphia: The American Philosophical Society; 1990. 164 p.

2. O'Malley CD. Berengario Da Carpi Giacomo 2008 [cited 2016 31 July]. Available from: http://www.encyclopedia.com.

3. Zimmerman L, Veith I. *Great wound surgeons; Hans von Gersdorff. Great Ideas in the History of Surgery*. New York: Dover Publications; 1967:203−217.

4. von Gersdorff H. *Feldbuch der Wundarznei*. Darmstadt: Wissenschaffliche Buchgesellschaft; 1517.

5. Ganz J, Arndt J. A history of depressed skull fractures from ancient times to 1800. *J Hist Neurosci*. 2014;2014(3):233–251.

6. Singer C. *Vesalius on the Human Brain*. London: Oxford University Press; 1952.

7. Majno G. *The Iatros. The Healing Hand*. Cambridge, MA: Harvard University Press; 1965:141–206.

8. Donaldson I. Ambroise Paré's accounts of new methods for treating gunshot wounds and burns 2004. Available from: http://www.jameslindlibrary.org/articles/ambroise-pares-accounts-of-new-methods-for-treatinggunshot-wounds-and-burns.

9. Hippocrates. Epidemics Book 5. Hippocrates. Cambridge, Massachusetts: The Loeb Classical Library, Harvard University Press, 1994, p. 179.

10. Keynes G. *The Apologie and Treatise of Ambroise Paré*. Chicago: University of Chicago Press; 1952.

11. Lowe P. The tenth chapter, of woundes in the head. The Whole Course of Chirurgie. London: Thomas Purfoot; 1612.

CHAPTER 6

The 17th Century

INTRODUCTION

The 17th century was a period of sociopolitical turmoil. In Britain, the Scottish Stuart family had taken over the English throne and in various ways demonstrated their failure to understand the times in which they lived. There was strife between the Protestant and Catholic faiths. Moreover, there was even more extreme strife between the various branches of the Protestant faith leading to the Civil War and the execution in 1649 of King Charles I. There were substantial advances made in a number of sciences and indeed in the nature of science itself. However, as will be seen later, the century was more important for men who would influence the future rather than men who made their greatest impact during their own life times. Maybe this is a reflection of the instability of the times. This was a time of preparation for the future and in consequence this chapter may be seen as a short digression from the main text, providing a link between the preceding and succeeding centuries.

RENÉ DESCARTES

One of the major changes was the understanding of the importance of mathematics as the language of the physical world. Indeed, the very nature of the scientific method was developing. In this process, the philosopher René Descartes (1596−1650) in his book a "Discourse on Method" laid down the four principles of scientific examination which may be summarized as follows.

1. Accept Nothing as True before excluding all ground for doubt.
2. Divide Difficulties into as many parts as possible.
3. Proceed from the Simple to the Complex.
4. Make Enumerations Complete and Reviews General.[1]

Intracranial Epidural Bleeding. DOI: https://doi.org/10.1016/B978-0-12-812159-7.00006-0

The substitution of accepted classical learning with observation and mathematical reasoning was not an approach appreciated by the authority of the Church, which made difficulties for a number of the earliest scientists. For example, Johannes Kepler (1571–1630) who initiated the mathematical approach to planetary movement suffered the disapproval of the Catholic Church and the ridicule of Martin Luther. However, as he lived in a Protestant part of Europe he suffered no more than inconvenience. Galileo Galilei (1564–1642) living in Catholic Italy had a much tougher time with the church, ending in his banishment from Florence to his estate outside that city where he lived out the rest of his life under house arrest. The study of reality with scepsis became the fashion and has remained so to the present day. Publishing the findings of such studies was an often socially dangerous adventure. The greatest achievement in the physical sciences and mathematics was the work of Sir Isaac Newton (1643–1727), the son of a Lincolnshire yeoman. His findings would affect all future scientific endeavors including medically related knowledge. He was made a Fellow of Trinity College Cambridge, an appointment that carried with it the necessity of becoming ordained into the church. Newton's extensive studies soon led him to views which were not in keeping with church authority. Through the intervention of friends, it was made possible for him to avoid ordination. However, this is an indication of how even Newton was not immune from theological orthodoxy, which necessitated him keeping his views quiet. Happily, he had a secretive disposition.[2]

WILLIAM HARVEY

Revolutionary in qualitative biological science was the publication of "Du Moto Cordis" in 1628 by William Harvey (1578–1657) the son of a prosperous Kent Yeoman. He had a long academic career studying in Cambridge and Padua. His Cambridge college was Gonville and Gaius which together with Padua provided a link with Vesalius, who had shared rooms with John Gaius in Padua during the previous century. A striking feature of Harvey's work is that it was achieved using ligatures and knives; equipment that was readily available all the way back to the classical world. During countless vivisection experiments, undertaken in a back room in his home, he either occluded vessels or incised them. What was new was the application of accurate observation to what occurred in his experiments.

Harvey's magnum opus was published in 1628 in Frankfurt by an English publisher called Fitzer. It was printed on poor quality paper, with cheap binding and contained 126 errors.[3] The book was met by widespread opposition from distinguished colleagues. The medical authorities of the day, brought up in the teachings of Galen, were not about to change their minds. Harvey was labeled a fantasist and his new ideas rejected. In this case, the opposition was not from the religious authority of a scandalized church but from the intellectual rigidity of a conservative profession. Nonetheless, as is ever the way, given time its truths came to be seen as irrefutable and it represented a major step forward in the application of science to medicine.

THOMAS WILLIS

Thomas Willis (1621−75) came from a family located somewhere within Berkshire and Oxfordshire and attended school in Oxford. He matriculated at Christ Church Oxford in 1638 obtaining a BA in 1639 and his MA in 1642 just before the outbreak of the Civil War. He served on the Royalist side in a University Regiment between 1644 and 46. He became interested in medicine and was awarded a BM in 1646 on the recommendation of the Royalist Regius Professor of Physick.[4] He had a long and distinguished career at Oxford undaunted by the restrictions of the Commonwealth. He had some talented and celebrated friends including the philosopher John Locke and the architect Christopher Wren who would illustrate the manuscript of Willis' most celebrated work "Cerebri Anatome." His eponymous arterial circle is illustrated in the woodcut from his book (Fig. 6.1). In addition to the arterial circle he is said to be the first person to have used the term "neurology."[5,6] Willis also suggested that the memory, will, and imagination were lodged in the cerebral parenchyma and not the ventricles, so he was still influenced by Galen's physiology. On the other hand, he also introduced the terms hemisphere, lobe, pyramid, corpus striatum, and peduncle.[6]

JOHN WOODALL

John Woodall (1570−1643) began his career at a time when surgeons had a much inferior position to physicians and he put great effort into combatting the perceived disadvantages of this unequal relationship. In time, he became a senior naval surgeon and clearly understood the

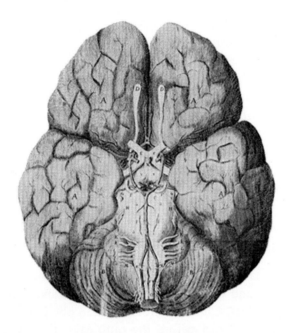

Figure 6.1 This engraving by Christopher Wren is far superior to that in Vesalius' work and anatomically much more accurate with a better depiction of the pons and cranial nerves. However, its major new material is the visualization of the anatomical arterial arrangement now universally known as the Circle of Willis.

principles of his craft. He was also very much an administrator and a bureaucrat. He spent much of his life fighting for so simple a matter as the right for a ship's surgeon to decide the contents of his surgeon's chest. These contents defined the equipment the surgeon had at his disposal.[7] He became Surgeon General to the East India Company on which in time much of Britain's Imperial wealth would depend. This position was being made untenable because of interference from the socially superior physicians who wished to have a say in the contents of these chests. After a lot of squabbling, Woodall won the conflict and was empowered to decide with a specific surgeon the contents of his chest. He also improved the quality of examinations for those wishing to become surgeons. Even more he was the first to stress the use of citrus fruits against scurvy. In contrast with frequent modern assumptions, he was also passionate about surgical hygiene writing: "Only in conclusion note, that it is very fit and needfull for the Surgion to have at the least two incision knives, one greater, one lesse, and that he keepe them sharpe and cleane; but let them not be so thinne grownde in the edge as the Rasor, for then they will deceive the workeman, when hee hath most use of them." He further states the following

Figure 6.2 This is similar to the trephine used by John Woodall who is said to have introduced the conical shape to the bit. From John Woodall's *The Surgeon's Mate.*

concerning trepans. "First be sure the instrument of it selfe be good, and of the best making, and that it be cleane from rust, and perfect without faults; for those Trapans which are brought from Germany are not to be used, nor yet to be tolerated."[8] How interesting to learn that metal engineering in Germany was regarded as inferior. The above texts are an indication that the common view that prior to Lister and Semmelweis all surgeons were filthy may perhaps not be true. Woodall qualified as a barber surgeon around 1600 and died in 1643 so that most of his important work was carried out in the reign of Charles I and before the Civil War.

There is evidence that Woodall was the first to use a trephine with a conical shape, to prevent the instrument sliding in and damaging the brain. Fig. 6.2 shows an instrument similar to that used by Woodall.[9]

RICHARD WISEMAN (1621–76)

Richard Wiseman[10] unlike Woodall was deeply involved in the Civil War as a royalist and suffered repeated imprisonment. Eventually, his loyalty was rewarded and he ended up Sergeant Surgeon to Charles II, a position equivalent to Surgeon General elsewhere. Little is known with certainty of his early life. However, during his career he collected and published reports of over 600 patients with such lucidity that Dr. Johnson praised his use of language. He had a scientific approach to surgery unusual for his time.[11] He raised the social perception of surgery and began the process which evolved into a situation where physicians were no longer seen as superior.

He continued active patient care even when sick. Thus, he was trepanning a patient on one occasion when he had to withdraw for a short time to cough up blood. Yet the next day he returned to dress the patient.[12] His management of the scalp was sound. He advised that it may be sutured which was not in keeping with contemporary practice. He opposed the use of a circular incision to lay bare the scalp when the only purpose was investigative for seeking a fracture. Instead he advised the cruciate or "T" shaped incision which can be closed. He also where possible avoided the then popular placing of dressings under the wound edges to keep the wound open and allow healing by secondary intention. He thought primary intention was preferable.

With respect to cranial trauma he bases his initial diagnosis on the same findings in the history as detailed by Hippocrates and Celsus. He believed incorrectly like so many of his predecessors that the pericranium arises from the dura through the sutures. He lays special emphasis on concussion, believing like his predecessors that it may be due to rupture of blood vessels. However, he distinguishes between minor concussion in which the victim is merely stunned and major concussion which may be as dangerous if not more dangerous than a fracture. This is a movement in the direction of the brain as a source of symptoms but he did not believe it was for reasons given later. He believed if there was a paralysis it indicated the membranes of the brain were damaged. He stated that if opening the skull did not relieve symptoms then there may be blood under the dura. He is clear over the fact that damage to the dura is an indication of a more severe injury.

Like all his predecessors he observed partial thickness fractures and bone contusions. On the other hand, his major indication for surgical intervention was concussion or a disturbance of consciousness. He believed if the dura was not relieved of blood collections there would be "Inflammation, Fever and Delirium; from whence Coma Convulsions Palsies and Death ensue, if neglected." The intervention could be with a trepan. However, if there was a fissure wide enough to permit drainage a trepan would not be necessary. He advised there was no reason to remove more bone fragments than necessary. He further advised the use of trepanation even in the absence of a fracture if there was coma or paralysis. If these dysfunctions were not relieved by exposing the dura, then that membrane must be opened because the blood may be between the dura and the pia or even in the ventricles. This is the first mention of this possibility. The reason for removing hematomas is that if left in place they become gangrenous (sphacelate) and the patient would die in coma with convulsions.

Wiseman saw a lot of penetrating injuries because he frequently attended men in battle. A consequence of this experience was that he came across and indeed handled injured cerebral tissue which he found to be insensible. He maintained the symptoms of convulsions and other symptoms as described earlier, which he calls accidents, are dependent on the corruption of the membranes. This is more or less in keeping with the teachings of Celsus. A minor detail is that Wiseman preferred the trepan (the brace and bit used today requiring two hands) to the trephine which could be rotated with a single hand.

CONCLUSIONS

The 17th century was a time of social and intellectual turmoil. Much of the efforts of this century would bear fruit in time yet to come. As has been indicated, new ideas take time to come to fruition. It is remarkable that in the face of passionate authoritative variants of Christian teaching, the scientific pioneers still managed, even when their liberty was at stake to continue their work and get it published. Two great surgeons contributed to the profession in general rather than to head injury management in particular. The first James Woodall worked to advance the reputation of the profession. What is important is his insistence on the meticulous cleanliness of surgical instruments, which is not at all in keeping with many reports on the times. His main area of interest was the navy, but he also improved the level of education necessary to qualify as a surgeon.

In keeping with such a military century, Richard Wiseman, the other major surgeon was an army man. He produced a wonderful collection of case reports on over 600 patients. Of these a number had head injuries. His reporting was systematic. He believed collections of blood were associated with coma, convulsions palsies, and death. However, his explanation for this was that the membranes around the brain were made rotten and corrupted. He believed the brain without sensation as illustrated in penetrating war wounds could not produce the symptoms. Thus, like so many of the historically important surgeons in this book, his observations were often sound but his explanations limited by the intellectual climate of his time. He was like so many others convinced that there were partial thickness skull fractures.

REFERENCES

1. Déscartes R. *A Discourse on Method*. London: Everyman; 1912.

2. Westfall RS. *Newton, Sir Isaac (1642–1727)*. Oxford: Oxford University Press; 2004. Available from: http://www.oxforddnb.com/view/article/20059.

3. Wright T. *Circulation: William Harvey's Revolutionary Idea*. London: Chatto and Windus; 2012.

4. Martensen RL. *Willis, Thomas (1621–1675)*. Oxford: Oxford University Press; 2007. Available from: http://www.oxforddnb.com/view/article/29587.

5. Birkwood K. Thomas Willis: the father of neurology London: Royal College of Physicians; 2016. Available from: https://www.rcplondon.ac.uk/news/thomas-willis-father-neurology.

6. Finger S. Changing concepts of brain function. In: Finger S, ed. *Origins of Neuroscience: A History of Explorations Into Brain Function*. Oxford: Oxford University Press; 1994:18–31.

7. Appelby JH. New light on John Woodall, surgeon and adventurer. *Med Hist*. 1981;25:251–268.

8. Woodall J. *The Surgions Mate*. London: Laurence Lisle; 1617.

9. KIrkup J. The evolution of cranial saws and related instruments. In: Arnott R, Finger S, Smith CUM, Lichterman LB, eds. *Trepanation: History, Discovery, Theory*. Lisse, The Netherlands: Swets & Zeitlinger B.V; 2003:288–300.

10. Wiseman R. *Eight Chirurgical Treatises*. London: Walthoe, J et al; 1734.

11. Kirkup J. *Wiseman, Richard (bap.1620, d. 1676)*. Oxford: Oxford University Press; 2008. Available from: http://www.oxforddnb.com/view/article/29792.

12. Smith AD. Richard Wiseman: his contributions to english surgery. *Bull NY Acad Med*. 1970;46(3):167–182.

CHAPTER 7

The 18th Century

INTRODUCTION

The previous chapters have outlined the principles on which the management of cranial trauma had evolved. The physiology of humors meant that many patients would have blood removed to improve the balance of these humors or in lesser cases would be purged. These practices would continue until the end of the 19th century. From ancient times, up to the Renaissance, the Church had resisted the introduction of new knowledge in all areas of science including medicine, since the teachings of Jesus were much more important than the findings of man. Nevertheless, as outlined in Chapters 4, From Ancient Times to the 17th Century, and 5, The Renaissance, there were a number of useful scientific observations made over the centuries including on the effects of cranial trauma on brain function. Nonetheless, none of them passed into the general corpus of medical practice. It was not until the 18th century that the role of the brain in cranial trauma became acknowledged.

The growing understanding of the role of the brain in the clinical picture following cranial trauma led to the gradual development of more rational head injury treatment. This was also the century in which writings about the practice of trepanation began to be clarified. There were altogether 11 great surgeons whose texts on the subject of cranial trauma gradually changed the face of neurosurgical practice. However, this understanding did not come at once but evolved and became more nuanced over the course of the century. The story of this evolution is the subject of this chapter.

DANIEL TURNER

Daniel Turner (1667—1741) was the son of a London candle maker who underwent a full apprenticeship to the Company of Barber Surgeons. He qualified in 1691. In 1711 he obtained a License from

Intracranial Epidural Bleeding. DOI: https://doi.org/10.1016/B978-0-12-812159-7.00007-2

the College of Physicians, which would have enhanced his status and speaks well of his reputation.[1]

Turner was the first 18th century author to publish on the topic of cranial trauma; the first issue of his work coming out in 1722. He distinguished three types of fracture, fissure, depressed and arched; the latter referring to a part of the bone being raised by the side of a part which is depressed. He was, like Paul of Aegina skeptical to the existence of contra-coup fractures unless the sutures were ossified and the whole skull was a rigid container. He subdivided fissures into fissures which are narrow and hard to detect and clefts which are wider. He instructed that wider clefts allow the passage of blood and matter relieving underlying structures. Fissures may require the use of the trepan. His approach to trepanation was as follows. "... if your Patient loose his Senfes, raves or utters any incoherent Words; also, if Vomiting, Palsy, Spasms or Convulsions; if Sopor or Snoring, as in a Lethargy or like one in a profound Sleep-, having taken away a convenient quantity of Blood. from his Neck, or, if that cannot be conveniently done, from either Arm, you are to cause his Head to be shaved, and then strictly examine all Parts of the Scalp."[2] Prior to this passage he has stated that the above-mentioned symptoms may be associated with either fracture, or concussion with associated effusion of blood upon the surface of the brain.

If the examination required earlier gave no result, then other people around should be questioned as to what hit the head and where on the head the blow landed. Having determined the definite or probable location of the trauma a cruciate or other suitable incision should be made. If the next day the patient was no better, trepanation was indicated. There follows a lengthy passage on trepanation technique which he stated is safe even in children with thin skulls if it is performed correctly. He wisely suggested that the trepanation disk should not be sawn right through. Instead when it could be felt that the trephined disk could move, drilling should cease. An elevator should be inserted underneath the disk at some point where the drilling had penetrated the full thickness of the bone. Then the remaining in-tact bone could be fractured by elevating the disk and avoiding damage to the underlying dura. He was thus limiting the indication for trepanation to patents with symptoms of cerebral compression. In addition, it was sometimes to be used in cases of depressed fractures. He like a gradually growing number of colleagues advised that elevating of all depressed fragments was not necessary and as many as possible should not be touched.

On the matter of cerebral function, he was revolutionary. He states "Till the business of Sensation is better unravelled, than hitherto it hath been, or I fear will be; the Disorders of the nervous system may be conjectured, but not clearly demonstrated. Whether there be any such Particles, as the animal Spirits is not yet universally agreed on." At the end of his chapter on cranial fracture he returns to this topic, partly reporting that the brain has no sensibility as Wiseman and others reported. However, he goes further. "That the Brain, justly supposed the Fountain as well of Motion as Sensation; is a Body of itself senseless; that, its Parts are not homogeneous; that its Substance is not strictly glandulous, nor properly medullary; that we may call it 'Substantia sui-ipsus, vel proprii generis,'[a] or, speaking plain truth, we may say. 'tis something, we know not what." Finally, he states "That our Knowledge, of the Causes and Effects of some nervous Distempers arrives but to a well grounded Guess, or rational Conjecture, and in likelihood will continue inexplicable to the human Understanding, under its present limited Conditions, as is the Essence of the Soul itself."[2] Thus, the knowledge of the brain is slowly beginning to be considered in more mechanistic terms and the notions of the humors are brought into doubt. This is a courageous text, opposing as it does over one and a half millennia of accepted doctrine.

Finally, on a more practical and less speculative note, Turner has the following sensible but macabre advice for the trainee surgeon. "But I shall no farther enlarge, by giving Rules for the setting on of the Instrument, unless this farther one, that no Person attempt to meddle therewith, who has not first well acquainted himself with the bony Compages,[b] and whole Encephalus, or the Structure of the Cranium, as well within as without; as also of the Meninges, and the Brain invested by them: And, for his farther Instruction, it may be necessary he frequently work with the Instrument upon humane Skulls, especially of Malefactors newly strangled, or others lately deceas'd, where such liberty maybe given without offence".[c2]

---

[a]A substance of its own nature.

[b]Compages = the entire structure (i.e., of the skull) as an integrated unit.

[c]Underlining by the author.

Turner is getting very close to applying his notion of brain function to his clinical assessment writing "As to the Prognosis, it is certain every fractur'd Skull has more or less Danger attending, not so much from the Fracture singly consider'd as the supervening Accidents by reason of the Membranes underneath oppress'd, Effusion of Blood upon the Brain, or this last itself, together affected." He was aware of brain pulsation stating for a woman who had lost a large amount of bone and is described thus "by which the Dura Mater lay expos'd for a vast compass, yet notwithstanding it incarn'd, though it never harden'd so but the Oscillatson, or alternate motion of the said Membrane, continu'd manifest." The lady was accustomed to wear a cap of beaten Lead and carried the large portion of her skull to receive alms. Again, in another place he refers to what we would call raised pressure with subdural "matter" in which case the "Membrane inflam'd lying high and turgid."

He specifically mentions cerebrospinal fluid which he calls lymph in the amount of "near a Pint." He is also aware of epidural blood. In another case when depressed fragments had been removed it is noted "the Dura Mater began to vibrate strongly."

Turner is most rational. He denies the importance of the phase of the moon, something which had been deemed to affect the position of the brain from classical times. He describes a sensible technique which is slower with less pressure in the thinner skulls of children. He propounds that if there is reason to undertake trepanation delaying it increases the risk. Unlike most of his countrymen he preferred the trepan to the trephine (Fig. 7.1).

LORENZ HEISTER

Lorenz Heister (1683–1758) was born in Frankfurt am Main, the son of an innkeeper and the daughter of a merchant. He had a distinguished education largely in the Netherlands with the celebrated clinician Herman Boerhaave (1668–1738).[3] His enormous "System of Surgery" was published in 1743, a work of unusual clarity of expression. He was the Professor of Surgery at Helmstadt and was also a Fellow of the Royal Society in London and of the Royal Academy in Paris. He was thus obviously a most respected surgeon in his own time. His surgery

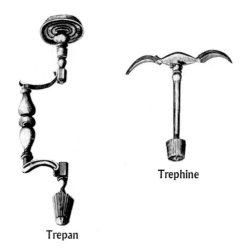

Trephine

Trepan

Figure 7.1 The two-handed trepan on the left was favored on the Continent and by some British surgeons. The one-handed Trephine was favored by most British surgeons. This image includes the same trephine showed in Fig. 6.2 plus an instrument from Heister's surgery text from 1743.

text includes sections on cranial injury, extravasation, and trepanation. There is much repetition of previously written notions but interestingly this is the first text to have a section specifically directed at the management of extravasations. He distinguishes between bleeding between skull and dura, between dura and pia, and under the pia. He specifies that deeper bleedings are more dangerous. He lists the usual set of symptoms which indicate that intracranial hemorrhage has occurred; "Restlessness, Delirium, Convulsions, Vertigo, Apoplexies, Stupidity, Loss of Senses, Speech and voluntary Motion." Heister mentions hemiparesis and quotes Morgagni in support of his contention that injury produces paralysis on the opposite side. He states "If either side of the Patient has lost Sense and Motion, and is become paralytic, it is an apparent sign, whatever some may think to the contrary, that the Injury was received on the contrary, or sound side. See Morgagni adversaria Anatomica'Ml. fc? Differt. de Refonitu Argentorat. 1722. Edit. Pag. 23."[4] He is the first to present this contention in respect of posttraumatic hemiparesis.[5] He also mentions a delay between injury and the development of symptoms as indication of accumulating blood pressing on the brain.

Heister finds himself on the horns of a dilemma. He knows that the rapid removal of extravasations should be done as quickly as possible. On the other hand, he is concerned that a trepan can injure the dura so that he is worried about unnecessary risk taken even though like Turner, he knew to stop drilling before the trepan has perforated all the way round and to use an elevator to free the still partially attached bone disc. Heister has one other technical contribution. He advised that if there was reason to believe there was intracranial hemorrhage and the first trepanation had no abnormal finding, he advocated multiple trepanations in a specific sequence first on the right and then on the left and from the front to the back of the head. He quotes Celsus as saying that in desperate cases it is much better to try a doubtful remedy than none at all. His basis for persisting with repeated trepanation was persisting bad symptoms as itemized earlier. He does *not* mention the state of the brain or the presence or absence of cerebral pulsation. He thus made substantial sensible contributions to the management of head injury many of which were ignored by his contemporaries or successors as will be seen in the ensuing paragraphs.

HENRI-FRANÇOIS LE DRAN

It is in the writings of the Académie Royale de Chirurgie in Paris, in a team led by Jean Louis Petit (1674−1750) that the modern truth gradually began to dawn. While Petit was the team leader, his work was not published until long after his death in 1774. It was thus his colleague Henri-François Le Dran (1685−1770) who brought the new ideas to a wider audience with a publication of his findings in English in 1742. Le Dran was also the son of a surgeon as revealed in his writings where he and his father manage a patient together. The London medical and scientific milieu was clearly impressed with him awarding him a Fellowship of the Royal Society on January 10, 1745. The crucial statement concerning the source of symptoms is based on an idea of almost alarming simplicity which required no new technology that was not available to Hippocrates. It goes as follows. "Must the Loss of Sense, which happened at the Moment of the Blow was given, be looked upon as a symptom of Fracture? NO because it continued only Half-a-quarter of an hour, the Fracture remaining in the same Condition for the space of a whole Day, without being dressed. Is it a Symptom of the Dura Mater's being lacerated? It is not, for the same Reason. Therefore, this Symptom can only be attributed to a Concussion of the Brain."[6]

The Paris school has also been celebrated for being the first to describe the lucid interval, so characteristic of epidural bleeding. However, careful analysis of their case material shows that this is not the case.[7] They did indeed described symptoms developing after a delay of some days but in Le Dran's material this delay was always associated with infection not bleeding. Nonetheless, this group also understood that the bad symptoms described by Heister earlier came from damaged brain. The group also realized that these symptoms could be due to both concussion and compression. The clinical comprehension of posttraumatic symptoms was made more difficult because the symptoms from concussion and compression were similar and could run into each other without an intervening interval. In Le Dran's cases with bleeding, there is no description of a period without symptoms between trauma and late deterioration. Rather there is a description of immediate symptoms running into deteriorating symptoms. Thus, Le Dran clarified the difference between immediate and delayed symptoms, but he did not describe a lucid interval.

Le Dran believed if the bone was contused and the pericranium had loosened following trauma this would be associated with an equivalent loosening of the dura from the inside of the cranium. He expressed this as follows: "If, in a contused Wound, where the Cranium is discovered, we find that the Pericranium adheres loosely to it, or is intirely separated from it, this is a certain Sign that the Cranium has suffered, although it is not fractured; and if that has suffered, we may justly conclude that the Dura Mater has suffered also. Hence, whensoever we find, by the Incision made, that the Pericranium has lost its Adhesion with the Cranium the Operation of the Trepan ought not to be deferred." The reason for urgent trepanation was the great risk in Le Dran's view of the development of epidural putrefaction.

Le Dran did make other contributions. He had an instinctive grasp of coping with raised intracranial pressure, including trying to maintain cerebral pulsation and shaving off cerebral hernias. He did not seem aware of lateralization of function or of contralateral hemipareses. He observed clinical deterioration with tight dressings without loosening them. Moreover, Le Dran was the first to describe a series of head injury patients with details of the individual management. However, he has the claim of the first properly documented successful surgical treatment of an epidural hematoma in 1708; his case XIX.[6]

Moreover, in his material there were altogether three epidural hematomas. One died from a coincident cerebral contusion. One died from a
subdural hematoma sustained following a fall while he was convalesc-
ing from his epidural bleed. Thus, it could be argued two operated
cases survived and the third died from a concomitant untreatable basal
lesion. These are very satisfactory results. All three patients were
operated within 3 days of the injury.

SAMUEL SHARP

Samuel Sharp (c. 1709–79) was born into a London family: his father
was a brass worker. He was lucky enough to be indentured to train
under the great surgeon Cheselden in 1725 at the age of 15. How this
was achieved is not clear. He also spent time studying in France after
he became acquainted with a French colleague who was visiting
Cheselden to learn his famous lithotomy technique. In 1733 at the age
of 24 he was appointed as a surgeon to Guy's Hospital. His "A
Treatise on the Operations of Surgery" work was first published in
1739 and the next year he got married. In 1746 William Hunter (the
less famous brother of John Hunter) took over from Sharp a series of
lectures at Covent Garden, on anatomy, surgery, and bandaging.
Sharp advised the composer Georg Friedrich Händel not to be operated for his cataracts. However, Händel was impatient and consulted
John Taylor (1703–72) another surgeon who had unsuccessfully operated on Johan Sebastian Bach. He did not help Händel.

Like Turner before him, Sharp considered that bad symptoms could
arise from fractures or concussions. However, he speculates on why some
concussions get better as follows. "In Concussions without a Fracture;
that produce the Symptoms here laid down, and do well afterwards, the
Vessels of the Brain and Membranes are only inflam'd and dilated; or if
they are ruptur'd, they absorb the extravasated Blood again."[8] He was
very keen that patients should be treated thoroughly with venesection
and enemas. He did not really make any other contribution but again
while his earlier explanation is speculative it is also mechanistic. He
unlike Turner preferred the single-handed trephine with a cylindrical
rather than a conical bit. He was firmly of the opinion that the risks of
trepanation were insignificant compared with the risks of failing to operate where surgery could help. He believed there was no reason to delay
surgery to obtain hemostasis. He argued for the use of scalping (removing

a more or less circular disk of skin) instead of a cruciate incision which could be closed. His arguments were related to the superior exposure so obtained. However, despite his eminence, Sharp's only real contribution was to encourage surgeons not to wait if they suspected trepanation could help the patient. There was no series of patients in his material.

JAMES HILL

James Hill (1703—76) was the eldest son of a Presbyterian Minister from a tiny village called Kirkpatrick-Durham a few miles from Dumfries. His surgical training was undertaken in Edinburgh. In 1772, 5 years after Pott's book on head injuries,[9] he published a book called "Cases in Surgery" which contained the case records of 18 patients who had suffered from cranial trauma.[10] Hill has since been forgotten, no doubt because he practiced in a small provincial Scottish town without the support of a large institution. In his own day, he was a much-respected practitioner throughout Great Britain. His management of patients with cranial trauma was unusually successful. Only 3 of his 18 patients died. Of these, two had untreatable cerebral contusions and one refused treatment.

Hill's contributions mainly relate to his superior understanding of intracranial physiology. However, his reporting introduced new methods even if the text could at times be a bit chaotic. For example, in 1772 he was reporting cancer cases using tables. His head injury cases had a mean follow-up of 17 years (range 3—30 years).

His clinical principles were biological rather than mechanical. More than any other 18th century author he was concerned with what we would consider intracranial pressure. He recorded cerebral pulsations in five out of nine patients he trepanned following trauma. He shaved cerebral hernias in three cases. He had no worry about incising the dura in marked contrast to Percival Pott. He learned that tight dressings could bring on the "bad" symptoms and he stopped using them. He was aware that neurological deficit was on the opposite side from the trauma and used it in his planning in two cases. This is unique in the 18th century. He treated four patients with epidural hematomas all of whom survived. Hill is not completely precise about the timing of surgery but it would appear that all operations were undertaken within the first 4 or 5 days after trauma.

He based his practice on his experience rather than on standard teaching, no doubt in part a result of his exposure as a young man to the evolving Scottish enlightenment. His results for the management of head injuries were the best in his century. His contribution will be easier to understand when recording the work of later colleagues who commented on his work. In this respect he recounts that he had accepted the teaching of many authorities including Turner and the great Cheselden mentioned earlier,[11] that the dura was attached loosely to the inside of the skull except at the sutures. However, a colleague he respected (he does not say whom but could it be Pott?) "flatly contradicted" this view. He discussed this with Alexander Monro primus who confirmed the newer view. Hill after some difficulty came on the basis of his own experience to change his mind and accept the uniform dense adherence of the dura to the inside of the skull. This is an example of his stated principle of preferring his experience to the teachings of authority. His results for the management of head injuries were the best in his century.

PERCIVAL POTT

Percival Pott (1714–88) came from an originally Cheshire family but his father was a lawyer in London and he was born in 1714 at what is now the site of the Bank of England. Pott's father died when he was 3 years old, but his widowed mother received financial assistance from the Bishop of Rochester. He thus underwent a quality education which culminated in his qualifying as a surgeon in 1736 at the tender age of 22. He went on to have a most distinguished career, a significant proportion of which was spent teaching. He published extensively, but sensibly instead of writing a single book on surgery he published shorter monographs on specific topics including one on head injuries.[12]

Pott's book on head injuries, published in 1768 was a turning point.[9] He reported 43 cases in systematic detail forming a wonderful source material for other surgeons. His statement of the cerebral origin of posttraumatic symptoms is even clearer than Le Dran's. "What are the symptoms of a fractured cranium? is often asked; and there is hardly any one who does not, from the authority of writers, both antient and modern, answer, vomiting, giddiness, loss of sense, speech, and voluntary motion This is the doctrine of Celsus." Pott goes on to state:

The symptoms just mentioned do indeed very frequently accompany a broken skull; but they are not produced by the breach made in the bone; nor do they indicate such breach to have been made. They proceed from an affection of the brain, or from injury done to some of the parts within the cranium, inde-pendant of any ill which the bones composing it may have sustained. They are occasioned by violence offered to the contents of the head in general; are quite independant on the mere breach made in the bone; and, either do, or do not accompany a fracture, as such fracture may happen to be or not to be complicated with such other ills.

They are frequently produced by extravasations of blood, or serum, upon, or between the membranes of the brain; or by shocks, or concussions of its sub-stance, in cases where the skull is perfectly intire and unhurt. On the other hand, the bones of the skull are sometimes cracked, broken, nay even depressed; and the patient suffers none of these symptoms. In short, as the breach made in the bone is not, nor can be, the cause of such complaints, they ought not to be attributed to it; and that for reasons, which are by no means merely speculative can only be the consequence of an affection of the brain as the common sensorium. They may be produced by its having been violently shaken, by a derangement of its medullary structure, or by unnatural pressure made by a fluid extravasated on its surface or within its ventricles.[9]

After this statement, it was generally accepted that the brain was the source of the bad symptoms and a great step was made to more rational treatment.

In addition, Pott had much to say on distinguishing a concussion from an extravasation although there are gaps in his reasoning. The symptoms concerned here are those quoted in the passage earlier. He considered concussion produced symptoms from shaking the brain but did not address the question of whether such loss of function is tempo-rary or permanent. With regard to the difference between symptoms from concussion and from extravasation, he quotes Le Dran as saying concussion symptoms come on early and extravasation symptoms come on after a latent interval. He then largely dismisses this pattern of events because he finds it rare in practice. The reason for this is that some extravasations occur at the time of injury. Some others produce symptoms early and merge into those of concussion. Thus, he finds timing of limited value when determining a course of action. In the event of a concussion with no external signs of trauma, surgery is pre-cluded for Pott because there is no means of knowing on which part of the cranium to operate. Where there are signs of trauma and bad symptoms trepanation is mandatory. If it is insufficient further trepanation may be considered at any location where there are indications of injury. He is very clear about the uncertainty of such surgery but argues with some force that there is no alternative and failure to proceed would constitute negligence. This is very much in keeping with the opinions of Samuel Sharp mentioned earlier.

Pott was also most concerned with the development of focal infections at the site of the trauma which came to bear his name as Pott's Puffy Tumor, a condition we no longer see. The symptoms concerned consist of headache, fever, local tender swelling with a later deterioration in conscious level. On exploration separation of the pericranium and intracranial purulent material was found.

This brings us to Pott's views on the anatomy of the dura and pericranium. He specifically states that the pericranium on the outer surface of the cranium and the dura on the inner surface are both densely adherent to the bone and function as nutritive periosteum. He states: "It has been thought by many, that the dura mater was attached to the skull, only at the futures; that in all other parts it was loose, and unconnected with it; and, that it constantly enjoyed or performed an oscillatory kind of motion; or was alternately, elevated and depressed. This idea, and opinion were borrowed from the appearance which the dura mater makes in a living subject, after a portion of the skull has been removed: but although it has been inculcate4 by writers of great eminence yet it has no foundation in truth or nature; and has misled many practitioners, in their opinions, not only of the structure and, disposition of this membrane, but in their ideas of its diseases."[9] This is the introduction to the modern concept that the dura is densely attached to the skull and this attachment will later be seen to be of great importance for the formation of epidural collections of blood.

While a huge contribution to the management of cranial trauma, the book was not without its eccentricities and weaknesses. Despite his understanding that the dura was attached evenly to the inside of the skull, Pott adhered to the same notion as Le Dran that loosening of the pericranium would also indicate loosening of the dura with the risk of the accumulation of infected material under the skull. Pott is also far more eager than many to elevate all fragments in a depressed fracture. He advised trepanation for fissures of the skull, but analysis of his cases shows that he did not practice what he preached.[13] He had no idea about treating what we call raised intracranial pressure so that there is no shaving of the brain, no mention of cerebral pulsation and no avoiding tight bandages (except for cranial fractures). He does not mention lateralization or neurological deficits contralateral to cerebral injury.

On the other hand, he does spend some time specifically on the topic of epidural hemorrhage. His advice seems curiously timid for one who was so willing to advise frequent use of the trepan. "Extravasations of any kind, and wherever situated within the cranium, are very hazardous, and much more frequently end fatally than happily; but considered as relative to the art of surgery, that which consists of merely fluid blood situated between the cranium and the dura mater is certainly the best, as it is nearest to the surface and admits the greatest probability of being relieved by perforation of the skull; grumous or coagulated blood, although in the same situation, by being most frequently adhering to the membrane, is not so readily discharged as the preceding, and therefore more likely to prove destructive: and all those which are either under the meninges, or within the cavities or substance of the brain, as they are very seldom within our exact knowledge, so they are also generally beyond the reach of our art."[9] There were three patients with epidural hematomas in Pott's material. Two were operated and survived. The other was not operated as there was no exterior traumatic mark to guide the correct location for trepanation. Both operated patients were operated within five days of the injury.

Pott is curiously hesitant when compared with other surgeons on the matter of incising the dura. He writes "The division of the dura is an operation which I have several times seen done by others, and have often done myself; I have seen it, and have found it now and then successful; and from those instances of success, am satisfied of the propriety and necessity of its being sometimes done: but let not the practitioner, who has not had frequent opportunity of seeing these kinds of things, presume from the light manner in which this necessary operation has been spoken by a few modern writers, that it is a thing of little consequence for it most certainly is not."[9]

In conclusion, whatever its weaknesses Pott's text by virtue of his reputation and clarity of expression was most influential in advancing the rational treatment of head injuries.

SYLVESTER O'HALLORAN

Sylvester O'Halloran (1728−1807)[14] was an experienced Limerick Surgeon. He was the son of well to do parents and his mother was related to a well-known Irish poet. He was educated in London, Leyden, and Paris. In 1793 he published a book on head injury management.[14] The text is opinionated and not always systematic but on the other hand his humanity shines through. He is emphatic that the connection between a loosened pericranium and loosened dura with concomitant risk, espoused by Pott was not in keeping with his observations. He states this as follows: "If the communication between the pericranium and dura-mater, by means of small blood-vessels, conveyed the injuries on one membrane to the other, why is it not constant and uniform, as we know Nature's laws invariably are? But this is by no means the cafe; and if researches were to be made, it would appear, that not one in twenty desperate wounds of the head, with denuded bone, have had any alarming symptoms whatever, during the progress of the cure! The advocates for this doctrine acknowledge, that numbers of cafes occur, in appearance very formidable, which are cured with little confinement or trouble."

Apart from providing a more accurate account of the behavior of the pericranium and dura, O'Halloran also treated three epidural hematomas. Two survived one died. The cause of death in that one case is not clear as the patient was doing well and then passed away. The three patients are however unusual because they were treated on, respectively, the 14th, 8th, and 12th days after trauma.

O'Halloran was not a supporter of the notion that there was a clinical picture which suggested treatable brain compression. On the contrary, he considered that concussion could explain all forms of disturbances of consciousness. If it were mild, the patient would wake up soon. If it were serious, then he/she would not. The location of the intracranial pathology which accompanied prolonged unconsciousness lay in the brain stem and was thus not accessible to surgery.

WILLIAM DEASE

William Dease (1752–98),[15] a Roman Catholic came from a prominent family. His father had lost his property for supporting the Stuarts at a time when the Hanoverians were on the English throne and the Stuarts were making ill judge attempts to get it back. He received his education in Dublin and Paris and settled in Dublin and achieved a distinguished career. In 1776 he published a book on the management of head injuries.[15] In the introduction, he is at pains to provide evidence that the practice advocated by Pott to trephine early if the pericranium is loosened to prevent epidural putrefaction is without purpose or justification. He had one patient with and epidural hematoma who was treated on the same day as his injury and who survived. The patient was struck on the head with a short sword but no details of the clinical course are available.

BENJAMIN BELL

Benjamin Bell (1749–1806)[16] was the son of a farmer in Canonbie in eastern Dumfriesshire. Bell had been trained in part by James Hill, of whom he had a high opinion. His surgical education was completed in Edinburgh. He wrote a surgical text book which included a long section on head injuries: though he did not present a clinical series. Thus, much of what he writes is more opinion than evidence though it reads well. The most worrying part is his rather unclear account of the indications for surgery. He lists up symptoms which may indicate concussion or compression. These are "Giddiness; dimness of sight; stupefaction; loss of voluntary motion; vomiting; an apoplectic stertor in the breathing; convulsive tremors in different muscles; a dilated state of the pupil, even when exposed to a clear light; paralysis of different parts, especially of the side of the body opposite to that part of the head which has been injured; involuntary evacuation of the urine and fæces; an oppressed, and in many cases an irregular, pulse."[16] That is fine but he is imprecise about the presence, persistence, and evolution of such symptoms. He protests about not using trepanation unless "bad symptoms" like those just quoted are present. In their absence, trepanation should be avoided. However, the imprecision of his account of the symptoms means that the text can all too easily be used as an excuse for unnecessary surgery. The absence of a case record does not help.

Bell's knowledge of cranial anatomy indicates that he knew the dura was firmly attached to the inside of the skull and the pericranium to the outside. He remains convinced that these attachments are markedly tighter at the sutures. He is further to be commended for getting the principles of the indications for trepanation right. He also specified contralateral hemiparesis in some cases. He also mentions fixed dilated pupils. This is more advanced than any predecessor. His technique is also of interest. He, in agreement with his successors prefers the trepan to the trephine, as it is quicker and gives better control. He advocates linear incisions not cruciate or scalping.

In cases where there are persistent symptoms of concussion/compression and no outer visible trauma he advocates multiple burr holes in a search for agents of cerebral compression. If the epidural space is clear and the dura is tense, then he advocates opening it. After surgery in the presence of raised intracranial pressure he advocated loose dressings.

Bell was a great professional success and it is interesting that he appreciates the adherence of the dura to the skull and the significance of contralateral pareses which is in keeping with the teaching of James Hill. To this Bell added fixed dilated pupils. There is however no mention of specific cases with epidural bleeding only general opinions about their management.

JOHN ABERNETHY

John Abernethy (1764−1831) the surgeon was the son of John Abernethy, a London Merchant according to two sources[17,18] and of a clergyman according to one.[19] His early training suffered from the help of his father, who arranged for him to be apprenticed to a friend and neighbor, one Sir Charles Blicks. There is agreement that Blicks was less than competent and that Abernethy probably learned much about how not to do surgery. More interesting, he seems to have been unenthusiastic about operating and is even recorded as not being the best of surgeons despite having a huge practice.

Abernethy reported three cases with epidural hematomas. They all died. However, it is fair to comment that two of the patients had very large hematomas and never woke up following the trauma. The other patient also had a large hematoma. However, he had been knocked unconscious and then woke up and walked home and went to bed. When the surgeon was called some hours later he was unconscious. This is a genuine lucid interval. The operation revealed a large hematoma and the brain did not re-expand following its removal. Abernethy referred to James Hill in his book and was clearly very impressed by him. Abernethy also was aware of contralateral paralyses as shown by the following quotation. "Should the state of the right side have been, as appears most probable, an approach to a state of paralysis, it must surely be considered as peculiarly curious. An effusion of blood in the left hemisphere of the brain would affect the opposite side of the body in the same manner, that cutting off the supply of blood to the left side appears in this instance to have done. I forbear to speculate on this subject: the fact which I have mentioned seems to deserve notice, and though at present it must stand alone, it may receive future confirmation, and when thus supported, be applied to the elucidation of physiology."[20]

CONCLUSION

Abernethy's book marks the end of the descriptions of head injuries in the 18th century. Clearly, there was patchy understanding of the management of raised intracranial pressure, crossed cerebral lateralization, and an approach which was based increasingly on experience rather than classical teaching. It was in the 19th century that these developments reached fruition with a clinical understanding of epidural hemorrhage which has changed little since that century.

REFERENCES

1. Wilson P. *Turner, Daniel (1667–1741)*. Oxford: Oxford University Press; 2008. Available from: http://www.oxforddnb.com/view/article/27844.

2. Turner D. *Of Fractures of the Cranium. The Art of Surgery. 1.* London: Rivington, C; 1736.

3. Ruisinger M. Heister, Lorenz. In: Bynum WBH, ed. *Dictionary of Medical Biography (Volume H-L)*. London: Greenwood Press; 2007:625–626.

4. Heister L. *Of the Principles of Wounds of the Head. A General History of Surgery in Three Parts.* London: Innys, W; 1743:82–92.

5. Schutta HS. Morgagni on Apolexy in De Sedibus: A Historical Perspective. *J Hist Neurosci.* 2009;18:1–24.

6. Le Dran H-L. *Observations in Surgery; Containing One Hundred and Fifteen Different Cases; With Particular Remarks on Each, for the Improvement of Young Students.* London: J Hodges; 1740.

7. Ganz JC. The lucid interval associated with epidural bleeding: evolving understanding. *J Neurosurg.* 2013;118(4):739–745.

8. Sharp S. *Of the Operation of the Trepan.* 3rd ed London: Brotherton J, Innys W, Manby R; 1740.

9. Pott P. *Observations on the Nature and Consequences of those Injuries to which the Head is liable from External Violence.* London: Lawes L, Clarke W, and Collins R; 1768.

10. Hill J. *Cases in Surgery.* Edinburgh: J Balfour; 1772.

11. Cheselden W. *The Anatomy of the Human Body.* London: Hitch and Dodsley; 1750.

12. Kirkup J. *Pott, Percival (1714–1788)*. Oxford: Oxford University Press; 2004. Available from: http://www.oxforddnb.com/view/article/22604.

13. Ganz J, Arndt J. A History of Depressed Skull Fractures from Ancient Times to 1800. *J Hist Neurosci*. 2014;(3):233–251. 2014.

14. O'Halloran S. *A New Treatise on the Different Disorders Arising from External Injuries of the Head*. Dublin: Z. Jackson; 1793.

15. Dease W. *Observations on Wounds of the Head*. London: G Robinson; 1776.

16. Bell B. *Of Affections of the Brain from External Violence. A System of Surgery*. Edinburgh: C Elliot; 1783.

17. Jacyna L. *Abernethy, John (1764–1831)*. Oxford: Oxford University Press; 2004. Available from: http://www.oxforddnb.com/view/article/49.

18. Waddington K. Abernethy, John. In: Bynum WBH, ed. *Dictionary of Medical Biography (Volume A-B)*. London: Greenwood Press; 2007:94–95.

19. Foster B. Sketch of John Abernethy. *Med Libr Hist J*. 1904;2(2):113–119.

20. Abernethy J. *Surgical Observations on Injuries of the Head*. London: Longman; 1810.

CHAPTER 8

The 19th Century
Evolution of Scientific Concepts

INTRODUCTION

The 19th century saw the final dawning of modern clinical understanding. All branches of modern pathology were dependent on the discovery of Cell Theory. Cells were first observed in plants by Robert Hook (1635–1703). He studied at Oxford with both Christopher Wren (1632–1723) and Thomas Willis (1621–75) of the Circle of Willis. The first to view motile cells was the Dutchman Antonie van Leeuwenhoek (1632–1723). The Cell Theory is credited to Mathias Jakob Schleiden (1804–81) who described cells in plants in 1838. In 1839 Theodor Schwann (1810–82) described cells in animals and together they propounded Cell Theory stating that all living things are composed of cells and cell products. However, the extension of this theory to the central nervous system (CNS) with the promulgation of the "Neuron Doctrine" would have to wait until the end of the century. Debate about whether the CNS was made of a network of fibers or interconnected individual cells had raged through the century.

CEREBRAL LOCALIZATION

While the contralateral localization of injury was finally being understood, another and more fundamental debate was born at the beginning of the 19th century. Franz Joseph Gall (1758–1828) developed a theoretical type of brain localization which he called phrenology. He localized functions or activities to areas that could be located on the skull. Not only did this nonsense lack any scientific foundation but the localization was designed in terms of specific functions like destructiveness or acquisitiveness (cerebral localization is in fact in terms of simple motor and sensory functions and not specific for given activities). In the event its theoretical weakness begat considerable

Intracranial Epidural Bleeding. DOI: https://doi.org/10.1016/B978-0-12-812159-7.00008-4

opposition and over time phrenology faded. However, while it was nonsense it stimulated the idea of cortical localization.[1]

This idea did not have an easy birth. There were significant scientists, particularly Friedrich Goltz (1834–1902), working in Strasbourg, who provided physiological evidence that cortical function was holistic not localized. This functioned as a counterbalance to the gradually accumulating evidence that cortical localization was a reality. The first step on the road to confirming localization was the work of Paul Broca who demonstrated the association between aphasia and a specific location in the left frontal lobe. This was followed by the work of John Hughlings Jackson (1835–1911), who described the cortical location of the origin of focal epileptic attacks which to this day still bear the name Jacksonian. His work was first published in 1864.[2] Despite these early stirrings there remained a school of thought that was firmly opposed to the notion of cerebral localization. The above-mentioned Goltz devised various experiments with dogs to confirm the truth of his opinions.[1]

On the other side of the debate were two Germans, Eduard Hitzig (1838–1907) a psychiatrist and Gustav Fritsch (1838–1927) an anatomist. Their experiments supported the notion of localized motor function in the brains of dogs. The leadership of the prolocalization movement was taken up by David Ferrier (1843–1928) a British scientist working with monkeys. The two sides in the debate came to a climax at the Seventh International Medical Congress in London in 1881. This was a truly amazing occasion which was opened in St. James' Hall, Piccadilly, in the presence of H.R.H. the Prince of Wales (the future King Edward VII), who was accompanied by his German cousin, His Imperial and Royal Highness the Crown Prince of Prussia (the future Frederick III) (see Fig. 8.1). A huge number of the contemporary medical science superstars were present. These included Sir James Paget, Lord Lister, Sir Jonathan Hutchinson, Rudolf Virchow, Louis Pasteur, Robert Koch, Jean Martin Charcot, and Sir William Osler. One American made a truly memorable and all too often forgotten recommendation. His name was General J.S. Billings and he presented four rules for the successful conference presentation.

1. Have something to say;
2. Say it;
3. Stop as soon as you have said it;
4. Give the paper a proper title!

Figure 8.1 A group photo taken at the 1881 Medical Exhibition in St. James Hall, Piccadilly.

How much better modern meetings would be if these simple rules were followed. However, we must return to the congress. Goltz presented dogs with extensive removal of both hemispheres without ensuing paralyses. On the other hand, Ferrier presented a monkey with a lesion in the motor cortex and a hemiparesis of which Charcot exclaimed "It is a patient." Subsequent examination revealed defects in Goltz's experimental technique revealing that his removal of brain tissue was in reality much less extensive than he had thought. That was the explanation for the discrepancy between his dogs and Ferrier's monkey. This really marks the beginning of the general acceptance of cortical localization which is important for this book.[1]

CLINICAL NEUROLOGY

The early 19th century was a period where the only guide to therapy was the clinical picture which was only then beginning to emerge from the mists of time. Surgeons were aware of the symptoms of brain injury which were classically vomiting, giddiness, depressed level of consciousness, loss of speech, and loss of voluntary motion. They had not been aware of the importance of hemiparesis, the state of the pupils, and the lucid interval. It would seem that James Hill and John

Abernethy were aware that a paralysis on one side meant a lesion on the opposite side.[3] Benjamin Bell mentioned fixed dilated pupils as did Abernethy who cited it as evidence of severe compression. He had no explanation for the phenomenon.[4]

JOHN BELL

It is said that John Bell (1763–1820)[5] was a compassionate man but he was certainly also passionate. It was as a result of his squabbles within the surgical fraternity in Edinburgh that George Bell, another brother, advised Charles to seek a career in London.[6] In his Principles of Surgery[7] John Bell criticized almost every known surgeon for over enthusiastic use of the trepan. He presented clinical presentations which in his mind permitted distinguishing between concussion and compression. Modern reality would consider his distinctions artificial and potentially misleading. However, they are delivered with such confidence that they could be persuasive. It could be argued that while there was a basis for his concerns he truly overstated his case. His influence was considerable and it is probable that his writings contributed to undue timidity in the use of skull surgery for the following century.

TREPANATION

During the first part of the 19th century, there was no consensus about the existence or clinical manifestations of epidural bleeding. Some surgeons had demonstrated that in their hands treatment could be successful but their message did not gain universal acceptance. It is a reasonable assumption that this was in part due to a lack of clarity over the nature of intracranial bleeding. It was also in part due to the passionate proselytizing against trepanation. While there was good sense in advising against the over use of this operation, the following section will show that it had a real value and the timidity of surgeons, without a proper theoretical basis for treatment contributed to the mortality of epidural bleeding.

1850s SIR JOHN ERICHSEN

John Erichsen (1818–96) was a highly successful Danish surgeon who practiced in University College Hospital in London. He was an unusual combination of professional excellence and personal affability.

One of his students was Lord Lister. He wrote a large text book on surgery which ran to many editions. The second was published in 1857 and devotes four pages to the subject of posttraumatic hemorrhagic intracranial extravasation, with particular emphasis on epidural bleeding.[8] He stated that the only certain clinical feature of extravasation was a progressively deteriorating level of consciousness. He described the clinical picture of initial unconsciousness, latent interval followed by increasing loss of consciousness which he attributed to what he called meningeal extravasation, due to a ruptured meningeal blood vessel. Hemiparesis or general paralysis would develop with dilated pupils. He also mentioned the variability of the clinical course. He did not mention the side of the dilated pupil. He emphasized a time frame of some hours. He suggested the treatment was trepanation but he said it was rare and recalled only a single case in 15 years. The progressive deterioration in conscious level without improvement was mostly likely due to cerebral extravasation where there was direct brain damage. This did not require surgery. He was thus in agreement with very limited use of trepanation, which he stated had been more popular previously. His reputation and the fact the text is found in a widely used book means that the views expressed would have been influential among his colleagues.

1867 SIR JONATHAN HUTCHINSON

It was in 1867 that Jonathan Hutchinson (1828–1913) finally wrote up the most important clinical characteristics of epidural bleeding. During the part of the century prior to his lectures there were somewhat vague and diverse concepts aired as outlined in the previous chapter. This must have been fairly confusing for the surgeons of the day. It would have been necessary to adopt one view or another of opposing notions since no consensus existed. The most important writers on epidural bleeding in the 18th century had not achieved any consensus either. Tables 8.1, 8.2 and 8.3 shows the results of the 18th century series in which epidural hematomas (EDH) were managed. In the first half of the 19th century the picture was however becoming clearer as indicated by the writings of John Erichsen mentioned earlier.

Hutchinson finally described the clinical picture in terms that are familiar today: "The importance of an interval of immunity between the accident and the occurrence of symptoms has long been recognized

Table 8.1 The Number of EDH Recorded in Clinical Series in the 18th Century

Author	No. of Head Injury Cases	No. of EDH Cases
Henri-François Le Dran[9] (1685–1770)	14	3
François Quesnay[10] (1694–1774)	39	4[a]
Percival. Pott[11] (1714–88)	43	3
James. Hill[12] (1703–76)	18	4
Sylvester. O'Halloran[13] (1728–1807)	71	3
William Dease[14] (1750–98)	24	2
John Abernethy[4] (1764–1831)	20	3

[a]Inadequate data to include in further analysis.

Table 8.2 The Different Surgical Notions of Six of the 18th-Century Writers

Author	Pericranial Separation Matters	Operate for Bad Symptoms	Elevate Depressed Fractures
Le Dran	Yes	Yes	Yes
Pott	Yes	Yes	Yes
Hill	No	Yes	Yes/no
O'Halloran	No	Sometimes	Yes
Dease	No	Yes	Yes
Abernethy	No	Yes	No

The Quesnay paper does not permit such an analysis. It can be seen that very different views were held. Pericranial separation refers to the idea that a cranial trauma can damage the blood vessels that run between the dura and the pericranium through the skull. They are considered to be the anchor of the membranes inside and outside the cranium. Damage to them permitted loosening of both membranes with the pericranium loosening being visible. The loosening of the dura was thought by Pott and Le Dran to encourage epidural putrefaction. The bad symptoms were stupor, a low intermitting pulse, nausea, vomiting, and sometimes convulsive twitches. Most surgeons took them seriously. However, O'Halloran believed that loss of consciousness was due to concussion which could be temporary or prolonged. He thus considered any prolonged bad symptoms could not be helped by surgery.

Table 8.3 The Relationship Between Injury and Contralateral Paresis

Author	Lateralization Understood
Le Dran	No
Pott	No
Hill	Yes
O'Halloran	No
Dease	No
Abernethy	Yes

as the chief indication of a ruptured meningeal artery; and it is to this, almost exclusively, that we must still give attention, if we wish to diagnose these cases." He later made the crucial quantitative contribution: The period at which the hemorrhage occurs may vary in these cases, just as it may after wound of an artery in one of the limbs. It is rarely profuse at first enough to cause symptoms. It may increase in the course of a few hours, or it may burst out suddenly, as a sort of secondary hemorrhage, at the end of 1 or 2 days afterwards. In rare instances, it may even be delayed for a week. He noted, in a fashion typical for neurological physicians that the results of surgery for EDH were dismal, "yet it is a remarkable fact that the modern annals of surgery do not, as far as I am aware, contain any cases in which life has been saved by trephining for such a state of things."[15]

This as we shall see was an inaccurate remark, despite the eminence and standing of Sir Jonathan. In his lectures Hutchinson suggested a mechanism for the fixed dilated pupil which was that it indicated compression of the oculomotor nerve.[15] However, since Hutchinson published in 1867 it is worth remembering the comment of B.M Patten: "... up to about 1850, clinicians were still trying to define and classify disorders of the nervous system. Little progress had been made and no advice could be given at the bedside of the paralyzed patient. Despite the advances in gross anatomy, pathology and physiology of the motor-sensory systems, no significant comparable clinical advances occurred."[16] Thus, Hutchinson's interpretation was a huge achievement in the absence of the techniques of neurological examination which are taken for granted today. He insisted that localization of a hematoma could be achieved by noting that a hemiparesis would be contralateral and a fixed dilated pupil would be ipsilateral. Moreover, there might often be a region of puffy swelling over the lesion.

He also made one other crucial contribution. He also pointed out that depressed fractures almost never produced signs of cerebral compression, thus ending a misapprehension that stretched back over centuries.

1867 AND LORD LISTER

The year 1867 was significant not only for Hutchinson's lectures, but was also the year Lister produced his first publication on antisepsis.

Judging from the case reports repeated in Jacobson's lengthy monograph on epidural bleeding these findings had little influence on the results of surgery for EDH throughout the century.

W.H.A. Jacobson's Monograph on Epidural Bleeding[17]

This text was and remains the definitive monograph on the clinical aspects of this topic. The text starts admirably insisting in keeping with Pott that the symptoms of compression from intracranial hematomas including those due to middle meningeal hemorrhage can mix with those from a primary brain injury and be indistinguishable. The EDH may exist on its own as well. What was definitely new was the acceptance of cerebral localization as demonstrated by David Ferrier. Jacobson's monograph includes diagrams taken from the papers of contemporary colleagues to illustrate the location of different cerebral functions. This is shown in Fig. 8.2. Thus, Jacobsen acquired the means whereby he could analyze the case reports which form the basis of his monograph. He also produced a wonderful woodcut which as well as any modern illustration shows the position and nature of an EDH produced by rupture of the middle meningeal artery. This is shown in Fig. 8.3. The course of this artery is shown in Fig. 8.4. These three figures illustrate the basic anatomical information which underpins the analysis of the 70 patients reported in Jacobson's monograph of them.

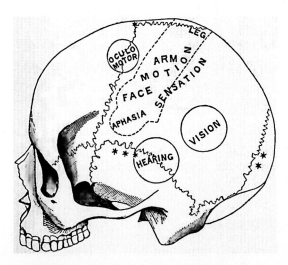

Figure 8.2 Cerebral localization as shown in W.H.A. Jacobson's manuscript on epidural bleeding.[17] The author is grateful to King's College London for permission to use this image.

The Lucid Interval

It was made clear that while Hutchinson had stated that a lucid interval was characteristic of epidural bleeding this was not born out by Jacobson's material. In the 63 cases where the clinical course is recounted inadequate detail, the lucid interval was clear cut and obvious in 32, probably present in 10 but could easily have been

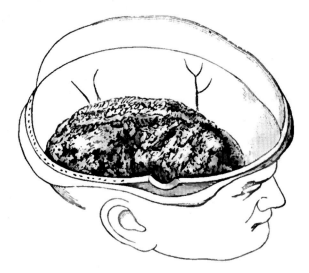

Figure 8.3 Location of an EDH due to middle meningeal rupture in W.H.A. Jacobson's manuscript on epidural bleeding.[17] The author is grateful to King's College London for permission to use this image.

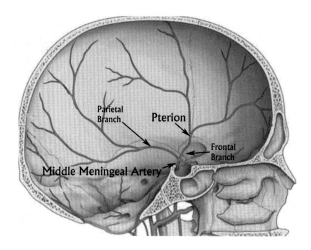

Figure 8.4 The course and major branches of the middle meningeal artery. At the pterion where the sphenoid, frontal, temporal, and parietal bones meet. The artery often runs in a bony canal in this location, which makes it easier for a fracture in the region to tear an opening in the artery.

overlooked. There were a further 21 where it was absent. The explanation was twofold. The interval might, as mentioned earlier, be absent because associated symptoms from brain damage could confuse the clinical picture. The other confusing factor familiar to all who manage head injuries in hospital was the consumption of alcohol. In this respect, nothing in our understanding of the clinical picture of epidural bleeding has changed since Jacobson's reporting.

Hemiparesis

The key finding was that hemiparesis when present was on the opposite side but it was by no means always present. Particularly in cases with an aggressive clinical course the patient becomes unconscious too quickly for a paresis to be observed. In the 70 cases, 32 had no mention of hemiparesis.

Pupils

Jacobson, influenced by Hutchinson was much impressed with the importance of the unilateral fixed dilated pupil. However, he makes the valid point that the finding is not limited to EDH but may be found in other types of cranial trauma. A fixed dilated pupil is mentioned in 23 cases. Of these 16 have the date of the trauma specified. Thus, 11 of the cases were reported 1–17 years after Hutchinson's paper which specified the significance of and mechanism underlying a fixed dilated pupil. In only one case in Jacobson's monograph, reported in 1881, 14 years after Hutchinson's lectures, was it mentioned that the author had not realized the significance of a fixed dilated pupil. That author was Jacobson himself. It would seem the surgeons exhibited innate conservatism and were also perhaps somewhat slow on the uptake. However, it must be born in mind there were no portable light sources for much of the 19th century. Thus, examining the pupil's response to light in a simple and reproducible way would have been difficult, especially for patients admitted after dark. It must also be remembered that electric lighting within buildings did not become generally available until the beginning of the 20th century. In respect of examination of the eyes, Helmholz introduced the first modern ophthalmoscope in 1851, but it was dependent on an external light source. The portable electric ophthalmoscope could not be available until the necessary batteries and bulbs were available, which also occurred at the start of the 20th century.

Infection

Lister's pioneering work on antisepsis was published, as mentioned in 1867, the same year as Hutchinson's paper on brain compression. It is surprising that only four patients died from wound-related infections. There were infections in a further three patients. In one case, there was erysipelas from a laceration in a nonoperated patient. The remaining six patients had all been operated. Apart from the four wound infections, in one case, there was a hyperpyrexia of unknown origin and in one case, there was pneumonia. That is of 23 patients treated with trepanation only 4 died from operation-related infections. That is only 17.4%. The one case in which Lister is mentioned had no wound infection but died of pneumonia. These findings must be placed in context. While there was lack of mention of antisepsis in the different reports, Jacobson in the introduction to the case reports in a discussion on the indications for early trepanation makes the following remark "While it is most right and necessary, especially in these days when antiseptic surgery has made trephining so much safer." Of the 4 patients who died from a wound infection, 1 occurred in the 10 patients operated before 1867. Three occurred in the 12 patients operated after 1867. Nobody doubts the importance of Lister's revolutionary improvement in surgical hygiene but the observations noted here cast some doubt about the significance of antisepsis in the management of this particular group of patients. Earlier work has suggested that the external milieu was a factor of major importance with regard to surgical infections.[18] This suggestion is confirmed by Jacobson's series in that all four deaths from wound infections occurred in London; a city with well-documented filthy conditions.[19,20]

Mortality Rate

Jacobson at the end of the monograph notes that he had followed a series of 70 patients of whom 10 survived. He mentioned a further 3 surviving patients but while that means he had a record of 13 surviving patients only 10 belonged to the series. That is a survival rate of 14.3%, which is not high. However, if we add the extra 3 surviving patients, not included in his series this gives 13 patients out of 73. To this may be added a further 11 patients because there are 3 EDH patients from Le Dran,[9] 3 from O'Halloran,[13] 2 from Dease,[14] 3 not 2 from Pott,[11] and 4 not 2 from Hill.[12] That gives a total of 84 cases of whom 21 survived. This gives an overall survival in 25% of patients. The survival rate for operated patients was 56.8%. This rather gives

Table 8.4 The Nature of Jacobson's Classification

Group	Characteristics
Mostly hopeful	Lesser violence. Fracture not involving skull base. Compression only. Direct brain injury absent or minimal
Less hopeful	Greater violence. Fracture may involve the skull base. Some direct brain injury but only trivial
Probably hopeless	Great violence. Extensive skull fractures. Severe direct brain injury

Table 8.5 The Frequency of Surgery, Survival, and Mortality Varied Between the Three Groups

	Mostly Hopeful	Less Hopeful	Probably Hopeless
Total	28	20	22
Surgery (trepanation/wound toilet)	16	5	7
Survived	10	0	0
Died from EDH/cerebral trauma	3	2	5
Died from neurological complications	2	0	0
Died from wound infection	1	1	2
Died from systemic infection	0	2	0

Patients with a favorable outlook had a greater chance of surviving. For the reasons for avoiding surgery, see Table 8.6.

the lie to Hutchinson's negative assessment of the value of surgery, and is a far higher figure than we customarily believe.

Jacobson classified his 70 cases. The classification is retrospective based on operation or postmortem findings. Table 8.4 shows that the management and mortality of the patients in the three groups differed.

Table 8.5 shows that 16 of the most hopeful category patients were operated with either wound toilet or trepanation. Ten of these survived which is a survival percentage of 62.5. This is estimable. Six died following surgery: three from the severity of the cranial injury, two from neurological complications, and one from a wound infection. Five of the less hopeful patients were operated. None survived. Two died from the severity of the trauma and three from infections, one in the wound, and two systemic. Seven of the probably hopeless cases were operated. None survived. Five died from the severity of the trauma and two from wound infections. These findings indicate that open surgery was of value in these cases especially in those which appeared to be less

Table 8.6 The Reasons for Not Operating in the Three Groups of Patients

Cause of Death	Mostly Hopeful	Less Hopeful	Probably Hopeless
Inadequate observation	5	3	0
No fracture, thus no action	2	0	1
Misinterpretation clinical findings	1	9	0
Speedy deterioration	0	3	4
Severe EDH and cerebral injury	0	0	8
Severe battlefield injury	0	0	1
Spinal injury	0	0	2
Total	8	15	16

Poor observation was by far the commonest cause of a treatable patient dying, while combined EDH and primary cerebral injury was the most frequent cause of death in the more severely injured patients.

severe. This conclusion is in contrast with the contemporary opinion outlined earlier.

Table 8.6 shows that failing to operate was in most instances probably unavoidable. In three cases, it was decided that there was no sign of a fracture which in keeping with the teachings of Pott meant no surgery was offered. In none of these cases was there a hemiparesis or unilateral pupil dilatation to guide the surgeon to a location for surgery. Thus, the absence of surgery is completely understandable. In the hopeless group 12 of 16 nonoperated patients either deteriorated too quickly for surgery or had such a severe injury that surgery would not have helped. There remain 18 patients who could have been operated in the other two groups. In 10 of these, the clinical picture was incorrectly interpreted since it would seem the clinical picture of possible epidural bleeding was not really understood. Then, eight suffered from inadequate observation so that the deterioration which could have led to surgery was missed. Cranial trauma observation, so familiar in the modern hospital was unknown. This was in all the cases involved due to ignorance not carelessness.

CONCLUSIONS

The 19th century saw the introduction of anesthesia and antisepsis which were the major new stepping stones away from suffering and toward safety. In the previous century, it had been finally determined that the brain was the source of symptoms following cranial trauma and not the cranium and meninges. This in turn led to the characterization of the clinical picture of falling level of consciousness, contralateral hemiparesis, and pupillary changes which are universally recognized today. On the other hand, the clinical picture remained the sole means of assessing a patient's diagnosis and a useful consistent way of examining the CNS had not been devised.

There was another important factor and that was the proselytizing of John Bell from Edinburgh. Though not as famous as his young brother Sir Charles, he inveighed against trepanation with passionate prose which often overstated his case. Conservatism concerning trepanation was also promulgated by such distinguished surgeons as John Erichsen and Jonathan Hutchinson. On the other hand, the analysis in this chapter documents the value of trepanation and its safety. Another surprising finding presented earlier is that the incidence of serious wound infections was so low. Maybe this complication has been overestimated.

The limitations of the clinical picture in determining diagnosis, the perceived need for caution with trepanation, and surgeons' innate conservatism prevented the development of any systematic pattern of management of the condition. Quite simply, there was a failure of awareness of how useful surgery was in the presence of epidural bleeding. This was compounded by a lack of awareness of the need for observing patients repeatedly following cranial trauma. This is hardly surprising when it is considered that while modern nursing is considered to have started in the middle of the 19th century, the establishment of professional bodies with state registration did not begin until the 20th. Thus, in 19th century hospitals there would have been nobody to undertake the necessary observations.

In the 20th century expanding clinical series and improving radiological techniques greatly improved the management of this eminently treatable condition. These included a period of observation of head injury patients. The technique of observation was a little varied until Teasdale and colleagues introduced the now invaluable Glasgow Coma Scale. These matters will be considered in more detail in the next chapter.

REFERENCES

1. Finger S. The era of cortical localization. In: Finger S, ed. *Origins of Neuroscience: A History of Explorations Into Brain Function*. Oxford: Oxford University Press; 1994:32–50.

2. Kerr P, Caputy A, Horwitz N. A history or cerebral localization. *Neurosurg Focus*. 2005;18(4):E1.

3. Ganz J. Head injuries in the 18th century: the management of the damaged brain. *Neurosurgery*. 2013;73(1):167–176.

4. Abernethy J. *Surgical Observations on Injuries of the Head*. London: Longman; 1810.

5. Baston KG. Bell, John (1763–1820) 2004 [cited 2016, 14 September]. Available from: http://www.oxforddnb.com/view/article/2013.

6. Gordon-Taylor G. The life and times of Sir Charles Bell. *Ann R Coll Surg Engl*. 1956;18:1–24.

7. Bell J. *The Principles of Surgery*. London: Thomas Tegg; 1826.

8. Erichsen J. *The Science and Art of Surgery*. 2nd ed London: Walton and Maberly; 1857.

9. Le Dran H-L. *Observations in Surgery; Containing One Hundred and Fifteen Different Cases; With Particular Remarks on Each, for the Improvement of Young Students*. London: J Hodges; 1740.

10. Quesnay F. Summary of observations on the use of the trepan. In: Ottley D, ed. *Observations on Surgical Diseases of the Head and Neck Selected from the Memoirs of the Royal Academy of Surgery in France*. London: The Sydenham Society; 1848.

11. Pott P. *Observations on the Nature and Consequences of those Injuries to which the Head is Liable from External Violence*. London: Lawes L, Clarke W, and Collins R; 1768.

12. Hill J. *Cases in Surgery*. Edinburgh: J Balfour; 1772.

13. O'Halloran S. *A New Treatise on the Different Disorders Arising from External Injuries of the Head*. Dublin: Z. Jackson; 1793.

14. Dease W. *Observations on Wounds of the Head*. London: G Robinson; 1776.

15. Hutchinson J. Four lectures on compression of the brain. *Clin Lect Reps London Hosp*. 1867;4:10–55.

16. Patten BM. The history of the neurological examination. Part 1: ancient and pre-modern history—3000 BC to AD 1850. *J Hist Neurosci*. 1992;1(1):3–14.

17. Jacobson WHA. On middle meningeal haemorrhage. *Guys Hosp Rep*. 1885/86;43:147–308.

18. Ganz J. Trepanation and surgical infection in the 18th century. *Acta Neurochir*. 2014;156(3):615–623.

19. Wallace AR. *The wonderful century it's successes and it's failures*. Toronto: Morang G.M; 1898.

20. Cockayne E. *Hubbub, Filth, Noise and Stench in England*. New Haven: Yale University Press; 2007.

CHAPTER 9

The 20th Century

INTRODUCTION

The work of W.H.A. Jacobson was and remains the classic text on the clinical picture following epidural bleeding. Little of principle has been added since. On the other hand, a variety of changes in healthcare have permitted the application of the information in Jacobson's monograph in practical clinical management. Nonetheless, while epidural hematoma management has improved greatly there are remaining difficulties to overcome. The evolution of head injury care in this chapter is limited to the United Kingdom because it is well documented and the necessary information is easily accessible to the author. It is believed that the changes in the United Kingdom were closely paralleled in Europe and North America although there would small variations related to locale and timing. The first half of the century saw only slow changes in head injury care which were not all related specifically to epidural bleeding. Nonetheless, the general improvement in head injury care and facilities for that care would also impact on the management of this specific form of hemorrhage.

1900—50

Neurological Examination

The development of the neurological examination was a great step forward. Sir Gordon Holmes was responsible for the examination introduced to the National Hospital for Nervous Diseases at Queen Square in London. This vital addition to the clinical armamentarium was finally published in 1946 but Holmes was on the staff of Queen Square from early in the 20th century and his teachings were adopted and much valued.[1] There are different variants of the examination with that used by surgeons was somewhat simpler. The examination permitted the recording or simple, reliable, reproducible objective findings. Changes in these can register a patient's course and indicate the seriousness of the situation.

Intracranial Epidural Bleeding. DOI: https://doi.org/10.1016/B978-0-12-812159-7.00009-6

Infrastructure

At the beginning of the 20th century, there was no comprehensive system of hospitals in the United Kingdom. This began to be rectified in the beginning of the 1930s.[2] Clearly, without easy access to hospitals systematic management would not be possible. The General Nursing Council (GNC), responsible for the state registration of nurses, was started in 1919. Thus, here is evidence of the gradual expansion of national health facilities before the Second World War. During that war a Head Injuries unit was established at St Hugh's College Oxford under the direction of Sir Hugh Cairns. It was so successful that it reduced the mortality from penetrating head injuries from 90% to 9%.

Instrumentation

The start of the century permitted greatly improved examination technology with the introduction of the ophthalmoscope and pocket torch. The otoscope had been around in one form or another since Guy de Chauliac (1300−68). The human electroencephalogram was introduced in 1924. Electromyography for humans was started in 1942.[3]

Images

While Wilhelm Konrad Röntgen (1845−1923) discovered X-rays in 1895 it took a while for this new method to become absorbed into every day clinical practice. Cushing obtained an early X-ray machine in 1896 and his first paper concerned the images of a bullet wound in the cervical spine. He reported this case based on X-rays just 2 days short of 1 year after Röntgen's discovery. This says much about Cushing's insight, determination, and efficiency. Even so there were limitations to the method. First, X-rays do not show soft tissues. Contrast media would be needed. The first used was air but in 1927 Egil Moniz introduced contrast materials based on halogens injected into blood vessels which demonstrated directly and indirectly a variety of soft tissue changes. The direct changes were diseases of the arteries themselves, including occlusions, malformations, and aneurysms. The indirect changes were displacement and stretching of arteries indicating a mass or space occupying lesion. This was for many years the definitive radiological investigation for epidural hematomas. Up to the 1930s diagnosis was naturally limited to the clinical picture. The introduction of carotid angiography permitted visualization of hematomas as shown in Fig. 9.1. Nonetheless, the establishment of radiology departments was a slow process only gathering speed in the 1930s.

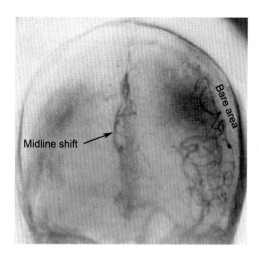

Figure 9.1 The image shows how the superficial cerebral vessels are pressed away from the inside of the skull by the hematoma. This produces the bare area. The mass effect of the hematoma not only presses the vessels immediately adjacent to it but also displaces the midline vessels across to the opposite side.

At that time, while the ability to make a clinical diagnosis was available together with a technique for visualizing hematomas on X-rays and while the profession of Nursing was well established there was still no all-embracing system of healthcare until after the Second World War. It is necessary to look at developments after that war to see how the modern management of epidural bleeding has come about.

There is one other factor to consider. If an epidural hematoma was suspected and no angiogram was available the surgeon could perform a frontal, parietal, and temporal burr-hole; a procedure given the informal name of wood-peckering. This could well reveal most sizeable hematomas. If no hematoma was found but the brain appeared tense through the burr-holes, a ventricle puncture could be performed. The finding of a dilated ventricle would justify the performance of posterior fossa burr-holes. By the same token difficulty in finding the ventricle would preclude investigating the posterior fossa. This method was in essence the same as that advocated by Lorenz Heister in the 18th century.

1951—2016

Treatment Delay—Getting the Patient to the Neurosurgeon

One of the most understandable areas of concern is the reduction of mortality rates. This is commonly seen to be related to delays in treatment.[4-10] There is general agreement that delay in treatment should be avoided. However, it would be impossible to draw up universally applicable guidelines. The situations in rural Australia, Norwegian valleys, Chinese heartlands, and the dense populations of the world's conurbations preclude such generalized solutions. It remains however the duty of those responsible for healthcare within their own communities to strive to set in place procedures which minimize delay. The pot of gold at the rainbow's end is a series of patients with zero mortality. This may be unachievable but it is a worthwhile aim for the future. The difficulty is that these procedures are costly in terms of staff and investment. Moreover, epidural hematomas are uncommon which does not help the prioritization of such procedures. It is not easy to find precise information on the incidence of epidural hematomas but one reliable source suggests 10% of serious head injuries may have this lesion. Certainly it is a minority.[11] It is suggested that while 1500 people per 100,000 attend Accident and Emergency department for head injuries on 15/100,000 require neurosurgical management.[12] Since epidural hematomas are only 10% of these injuries it is obvious they are not common. However, they are potentially treatable hence the dilemma on how to manage patients who might have the condition.[13]

This author spent 6 months in rural Norway as part of the requirement for obtaining a medical license. He saw no epidural bleeds during that time but head injuries were one of the most difficult cases to manage. As stated earlier only a minority will have an epidural but the observation of patients who have been unconscious but were now awake again is pretty much entirely aimed at catching this kind of hemorrhage just because it is treatable. In rural lightly populated Norway, the dilemma was the nearest hospital was across the fjord. To send a patient there required using a ship which meant waking quite a lot of people and at a considerable cost. Thus, it was usual to give instruction on observation to the family with the message to contact the doctor again if the patient deteriorated in any way. Luckily in my 6 months that never happened.

In the United Kingdom, as elsewhere, the statistics mentioned earlier indicate that there are far more head injuries presenting at Accident and Emergency departments than will require treatment. The capacity of Departments of Neurosurgery means that they will not be able to observe all the patients who have lost consciousness temporarily but are now fine. It is generally accepted that such patients need to be observed overnight in case they are harboring an epidural hematoma. In consequence, they are cared for in surgical (or occasional neurological) departments. It is not easy to ensure optimal observation in this context, no matter how willing the staff may be.

In the days before, modern computerized imaging the law courts placed a great deal of emphasis on skull X-rays. While these are of limited value in the management of head injuries, they do give an indication of the severity of the injuries. However, to be of value they must be taken correctly. This is more difficult than it sounds. The images should be taken from in front and from the side. The planes of the skull should be at a precise right angle to the incidence of the X-ray beam. It is hard enough to do this in the middle of the night but the difficulties are compounded in that the patients are often uncooperative because of the effects of either alcohol or of the head injury itself.

Treatment Delay—Useful Observation in Hospital

In the case of epidural bleeding, as mentioned earlier, the trick is to catch the condition early. By the time there is a marked hemiplegia and especially a fixed dilated pupil, the situation may be too late. In this context, it is the examination of the patient's mental function that is the key. In the late 1960s and early 1970s when this author was training, the literature once again gave expression to the conservatism of surgeons. Patients who had disturbed consciousness following cranial trauma were described among other things as sleepy, drowsy, stuporous, soporous, somnolent, and/or confused. These terms were not useful in determining the course of the patient's condition. However, this was all changed when level of consciousness charting was introduced using what has come to be known as the Glasgow Coma Scale (GCS),[5,14] as shown later.

Glasgow Coma Scale

Eye Opening	
Spontaneous	4
To speech	3
To stimulus	2
None	1

Verbal Response	
Orientated	5
Confused	4
Inappropriate words	3
Incomprehensible sounds	2
None	1
Motor Response (best)	
Obeys commands	6
Localizes stimulus	5
Flexion withdrawal	4
Flexion abnormal	3
Extension	2
No response	1

This simple scale is easy to operate. It requires minimal training. Nurses find it useful. It may be repeated every hour or less the first night after trauma. The result of any given examination results in a figure the minimum of which is 3 and the maximum is 15 and this number is charted. The charts are easy to follow clearly demonstrating improvement, deterioration or no change. This simple but brilliant tool replaced all the vague terminology. It records the patient's condition in terms not of words but of responses. It is used pretty much universally not only with epidural hematomas but with all forms of intracranial injury.

Computerized Imaging

In 1973−74 computerized tomography (CT) was introduced. This completely revolutionized the management of all head injuries including epidural hematomas. For the first time a blood clot could actually be visualized (see Fig. 9.2) Moreover, as will be indicated later, details of its formation could be shown in some patients. In addition, the CT of the skull was far better than a plane skull X-ray at demonstrating the details of a cranial fracture. Indeed, the details of skull base fractures, difficult to see on plane images can be easily shown on the CT. The introduction of magnetic resonance imaging a decade later had little significance for the assessment of patients in the acute stage but has been invaluable for demonstrating the long-term consequences of cerebral injury.

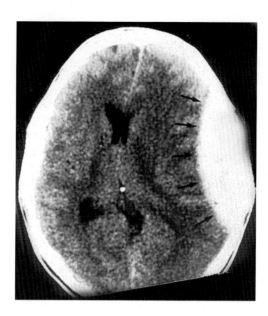

Figure 9.2 CT of an epidural hematoma. The arrows indicate the convex inner margin. Separating the dura from the skull results in the inner margin of the hematoma follows the separated dura and not the contour of the brain, thus distinguishing the lesion from a subdural hematoma. The ventricles are also shown displaced toward the opposite side.

Prognostic Factors in the Management of Epidural Hematomas

Speedy assessment and treatment is central to successful management.[4–10] One thing is very clear today and that is CT has greatly facilitated diagnosis and improved results. One series which exemplifies this consisted of 82 unselected patients. The mortality rate before CT was 29.2% and after it was 12.1%. Moreover, the morbidity sank from 31.7% to 19.5%.

Clinical factors which correlate with a worse outcome are a low GCS value at the time of surgery[15–20] and the presence of associated intracerebral lesions.[17,21–23] The lucid interval was an unreliable prognostic indicator.[6] This is not least because the presence or absence of this interval varies greatly between series and is by no means a constant feature of epidural bleeding. The negative significance of fixed dilated pupils was observed repeatedly.[6,19,21,22,24,25] Thus, as discovered in the 19th century, the classical clinical picture is still headache, contralateral hemiparesis, ipsilateral fixed dilated pupil, and a linear fracture. Nonetheless, none of these features is present in every patient making clinical diagnosis more difficult than had been previously thought.

Special attention was paid to the less common posterior fossa epidural hematoma.[25–35] The clinical picture is nonspecific characteristically typical of raised intracranial pressure with headache and sometimes ataxia. Clarity is obtained following CT.

Surgical Treatment[36]

Speedy removal of an epidural hematoma has been the keystone of treatment. Many authors recommend a craniotomy.[4,37,38] Initial decompression may be achieved through an initial burr-hole, restoring pulsation and thereafter a flap may be raised. An old-fashioned crown trephine is also a very efficient way of treating these lesions, though today it is out of fashion. The technique employed will be that which is approved on any given department. What is required is adequate exposure to permit removal of the hematoma and adequate access to stop any meningeal arterial bleeding. Sticking things into the foramen spinosum sounds sensible but has no place in real surgery. It should be mentioned that the introduction of blood and fluid transfusions and the development of plastic bags and tubes for the delivery of blood of fluids greatly increased the effectiveness and safety of surgical treatment.

Increasingly, toward the end of the 20th century, papers have been published providing evidence that conservative management is acceptable in some cases.[8,16,32,33,35–37,39–43] This has only become possible since CT has permitted the visualization of the intracranial status.

CONCLUSION

Thus, the 20th century added sophistication to the assessment of the clinical picture. It provided large series which form guidelines for surgeons. The imaging developed to the point that a definitive CT can be made available in 2–3 min. It has not however solved the central problem. Head injuries are common and epidural hematomas are rare. It is not feasible at present for every head injury patient to have a CT since the cost would be great and moreover the capacity does not exist. Because of the variability of the clinical picture there is still no simple procedure for consistently choosing the patients who require a CT. Thus, the while the dilemma associated with these hematomas has been combatted quite effectively it has not gone away completely. Improvement is still possible.

REFERENCES

1. Holmes G. *Introduction to Clinical Neurology*. London: E&S Livingstone; 1946.

2. Carruthers GB, Carruthers LA. *A History of Britain's Hospitals*. Sussex, England: Book Guild Publishing; 2005.

3. Kazame M, Province P, Alsharabati M, Oh S. History of electromyography (EMG) and nerve conduction studies (NCS): a tribute to the founding fathers. *Neurology*. 2013;80 (7Supplement P05.259).

4. Choux M, Grisoli F, Peragut JC. Extradural hematomas in children. 104 cases. *Childs Brain*. 1975;1(6):337–347.

5. Teasdale G, Galbraith S, Murray L, Ward P, Gentleman D, McKean M. Management of traumatic intracranial haematoma. *Br Med J (Clin ResEd)*. 1982;285(6356):1695–1697.

6. Dan NG, Berry G, Kwok B, et al. Experience with extradural haematomas in New South Wales. *Aust N Z J Surg*. 1986;56(7):535–541.

7. O'Sullivan MG, Gray WP, Buckley TF. Extradural haematoma in the Irish Republic: an analysis of 82 cases with emphasis on 'delay'. *Br J Surg*. 1990;77(12):1391–1394.

8. Browne GJ, Lam LT. Isolated extradural hematoma in children presenting to an emergency department in Australia. *Pediatr Emerg Care*. 2002;18(2):86–90.

9. Deverill J, Aitken LM. Treatment of extradural haemorrhage in Queensland: interhospital transfer, preoperative delay and clinical outcome. *Emerg Med Australas*. 2007;19 (4):325–332.

10. Bulters D, Belli A. A prospective study of the time to evacuate acute subdural and extradural haematomas. *Anaesthesia*. 2009;64(3):277–281.

11. Hilton DA. The neuropathology of head injury. In: Whitfield PC, Thomas EO, Summers F, Whyte M, Hutchinson PJ, eds. *Head Injury A Multidisciplinary Approach*. Cambridge: Cambridge University Press; 2009:13–21.

12. Samandouras G. *The Neurosurgeron's Handbook*. Oxford: Oxford University Press; 2010:208.

13. Phonprasert C, Suwanwela C, Hongsaprabhas C, Prichayudh P, O'Charoen S. Extradural hematoma: analysis of 138 cases. *J Trauma*. 1980;20(8):679–683.

14. Teasdale G, Murray G, Parker L, Jennett B. Adding up the Glasgow Coma Score. *Acta Neurochir Suppl (Wien)*. 1979;28(1):13–16.

15. Bricolo AP, Pasut LM. Extradural hematoma: toward zero mortality. A prospective study. *Neurosurgery*. 1984;14(1):8–12.

16. Servadei F, Piazza G, Seracchioli A, Acciarri N, Pozzati E, Gaist G. Extradural haematomas: an analysis of the changing characteristics of patients admitted from 1980 to 1986. Diagnostic and therapeutic implications in 158 cases. *Brain Inj*. 1988;2(2):87–100.

17. Palomeque Rico A, Costa Clara JM, Cambra Lasaosa FJ, Luaces Cubells C, Pons Odena M, Martin Rodrigo JM. Epidural hematoma in children. Prognostic factors. Analysis of 70 cases. *An Esp Pediatr*. 1997;47(5):489–492.

18. Ono J, Yamaura A, Kubota M, Okimura Y, Isobe K. Outcome prediction in severe head injury: analyses of clinical prognostic factors. *J Clin Neurosci*. 2001;8(2):120–123.

19. Cheung PS, Lam JM, Yeung JH, Graham CA, Rainer TH. Outcome of traumatic extradural haematoma in Hong Kong. *Injury*. 2007;38(1):76–80.

20. Fabbri A, Servadei F, Marchesini G, Stein SC, Vandelli A. Early predictors of unfavourable outcome in subjects with moderate head injury in the emergency department. *J Neurol Neurosurg Psychiatry*. 2008;79(5):567–573.

21. Jonker C, Oosterhuis HJ. Epidural haematoma. A retrospective study of 100 patients. *Clin Neurol Neurosurg*. 1975;78(4):233−245.

22. Lee EJ, Hung YC, Wang LC, Chung KC, Chen HH. Factors influencing the functional outcome of patients with acute epidural hematomas: analysis of 200 patients undergoing surgery. *J Trauma*. 1998;45(5):946−952.

23. Chowdhury SN, Islam KM, Mahmood E, Hossain SK. Extradural haematoma in children: surgical experiences and prospective analysis of 170 cases. *Turk Neurosurg*. 2012;22 (1):39−43.

24. Kvarnes TL, Trumpy JH. Extradural haematoma. Report of 132 cases. *Acta Neurochir (Wien)*. 1978;41(1-3):223−231.

25. Holzschuh M, Schuknecht B. Traumatic epidural haematomas of the posterior fossa: 20 new cases and a review of the literature since 1961. *Br J Neurosurg*. 1989;3(2):171−180.

26. Brambilla G, Rainoldi F, Gipponi D, Paoletti P. Extradural haematoma of the posterior fossa: a report of eight cases and a review of the literature. *Acta Neurochir (Wien)*. 1986;80 (1-2):24−29.

27. Pozzati E, Tognetti F. Spontaneous resolution of acute extradural hematoma—study of twenty-five selected cases. *Neurosurg Rev*. 1989;12(Suppl. 1):188−189.

28. Otsuka S, Nakatsu S, Matsumoto S, et al. Study on cases with posterior fossa epidural hematoma—clinical features and indications for operation. *Neurol Med Chir (Tokyo)*. 1990;30 (1):24−28.

29. Ciurea AV, Nuteanu L, Simionescu N, Georgescu S. Posterior fossa extradural hematomas in children: report of nine cases. *Childs Nerv Syst*. 1993;9(4):224−228.

30. Mahajan RK, Sharma BS, Khosla VK, Tewari MK, Mathuriya SN, Pathak A, et al. Posterior fossa extradural haematoma—experience of nineteen cases. *Ann Acad Med Singapore*. 1993;22(Suppl. 3):410−413.

31. Oliveira MA, Araujo JF, Balbo RJ. Extradural hematoma of the posterior fossa. Report of 7 cases. *Arq Neuropsiquiatr*. 1993;51(2):243−246.

32. Suyama Y, Kajikawa H, Yamamura K, Sumioka S, Kajikawa M. Acute epidural hematoma of posterior fossa: comparative analysis between 20 cases in adults and 10 cases in children. *No Shinkei Geka*. 1996;24(7):621−624.

33. Bozbuga M, Izgi N, Polat G, Gurel I. Posterior fossa epidural hematomas: observations on a series of 73 cases. *Neurosurg Rev*. 1999;22(1):34−40.

34. Dirim BV, Oruk C, Erdogan N, Gelal F, Uluc E. Traumatic posterior fossa hematomas. *Diagn Interv Radiol*. 2005;11(1):14−18.

35. Sencer A, Aras Y, Akcakaya MO, Goker B, Kiris T, Canbolat AT. Posterior fossa epidural hematomas in children: clinical experience with 40 cases. *J Neurosurg Pediatr*. 2012;9 (2):139−143.

36. Servadei F, Faccani G, Roccella P, et al. Asymptomatic extradural haematomas. Results of a multicenter study of 158 cases in minor head injury. *Acta Neurochir (Wien)*. 1989;96 (1-2):39−45.

37. Ciurea AV, Kapsalaki EZ, Coman TC, et al. Supratentorial epidural hematoma of traumatic etiology in infants. *Childs Nerv Syst*. 2007;23(3):335−341.

38. Taussky P, Widmer HR, Takala J, Fandino J. Outcome after acute traumatic subdural and epidural haematoma in Switzerland: a single-centre experience. *Swiss Med Wkly*. 2008;138 (19−20):281−285.

39. Pozzati E, Tognetti F. Spontaneous healing of acute extradural hematomas: study of twenty-two cases. *Neurosurgery*. 1986;18(6):696−700.

40. Chan KH, Mann KS, Yue CP, Fan YW, Cheung M. The significance of skull fracture in acute traumatic intracranial hematomas in adolescents: a prospective study. *J Neurosurg.* 1990;72(2):189−194.

41. Bezircioglu H, Ersahin Y, Demircivi F, Yurt I, Donertas K, Tektas S. Nonoperative treatment of acute extradural hematomas: analysis of 80 cases. *J Trauma.* 1996;41(4):696−698.

42. Heyman R, Heckly A, Magagi J, Pladys P, Hamlat A. Intracranial epidural hematoma in newborn infants: clinical study of 15 cases. *Neurosurgery.* 2005;57(5):924−929. discussion-9.

43. Teichert JH, Rosales Jr PR, Lopes PB, Eneas LV, da Rocha TS. Extradural hematoma in children: case series of 33 patients. *Pediatr Neurosurg.* 2012;48(4):216−220.

Intermediate Summary

CHAPTER 10

Historical Summary

INTRODUCTION

Both Hippocrates and Celsus mentioned lesions that sound like epidural hematomas. However, apart from advising that the blood should be allowed escape, they do not have an idea about different intracranial hematomas, their location, and effects. It was not until the 19th century that there was enough understanding of basic biological science that intracranial hematomas could be classified and characterized. Nonetheless, the basic science mentioned evolved over more than two millenia. It was not a straightforward process and for clarity, the different elements in it which are described in the previous chapters are summarized here, before proceeding to the next section concerning the pathophysiology of these bleeds.

CRANIAL ANATOMY

There was much written about calvarial anatomy and the anatomy of the sutures. The knowledge was sufficient to permit the development of techniques of cranial surgery including trepanation, wound toilet, and the elevation of depressed fractures. These techniques are still in use today with modern modifications and need concern us no further.

THE RELATIONSHIP BETWEEN THE SKULL AND THE DURA

Galen stated the dura was attached to the interior of the skull only at the sutures, because of a difference in the nature of the dura and the bone. A consequence of this error was that surgeons believed that they should not perform a trepanation which involved a suture since this could increase the risk of damage to the underlying dura. The error was repeated by many of the writers up the 19th century. They include William of Saliceto (1210−77), Guy de Chauliac (1300−68), Andreas Vesalius (1514−64), Ambroise Paré (1510−90), Richard Wiseman

Intracranial Epidural Bleeding. DOI: https://doi.org/10.1016/B978-0-12-812159-7.00010-2

(1621–76), Daniel Turner (1667–1741), and William Cheselden (1688–1752). Giacomo Berengario Da Carpi (1460–1530) alone correctly observed at an early date that the dura is densely attached all over the interior of the skull. The matter was finally laid to rest by Alexander Monro (1697–1767) and Percival Pott (1714–88). It is interesting and depressing that the correct anatomical relationships has been observed in the 16th century but that these observations were ignored.

FRACTURES

Cranial fractures were classified with a variety of different details but the classifications all come down to either fissure or depressed fracture. There was a fear that the injury resulting in the fracture could lead to putrefaction lying under the lesion with lethal results. Thus, if a fracture was narrow, it would be necessary to perform a trepanation to allow the escape of any accumulated matter and thus avoid the putrefaction. If a fissure was broad this would be enough to allow the matter to escape without any further intervention. It was widely believed that elevation of depressed fractures was indicated to relieve pressure on the brain. Celsus had maintained it was not necessary to elevate all depressed fractures but Galen was very keen on it. His view prevailed, not least because Celsus work disappeared for over a millennium. Even in the 18th century most surgeons were keen to elevate fractures. James Hill (1703–76) and John Abernethy (1764–1831) were the first to show it was not necessary in many cases. It was Sir Jonathan Hutchinson (1828–1913) who finally demonstrated that depressed fractures did not produce much in the way of cerebral symptoms.

One phenomenon which is unfamiliar in modern practice is the partial thickness fracture. It was Hippocrates who first described a technique using black dye to permit assessment of the depth of fissures. Galen went further and described in detail the best operative technique to clarify this issue, together with the instruments which would be required. Partial thickness fissures were mentioned by many writers down the centuries. They include Paul of Aegina (625–690), who is repeating the teachings of his predecessors. Roger Frugard of Parma (1140–95) introduced the use of the Valsalva maneuver to determine if a fissure was full thickness. If it were then the communication with the space within the cranium would be demonstrated by the flow of fluid out of the fissure. It would seem that Guy de Chauliac made similar observations. Wiseman also observed partial thickness fractures. The descriptions of the lesions, their diagnosis, and management were consistently reported over many centuries. It would be inappropriate to dismiss the existence of these lesions. It is possible that these were injuries inflicted by sharp weapons such as knives and swords, which are no longer in use.

CRANIAL CONTUSIONS

Today if MEDLINE is searched for bone contusion there are under 50 hits and most of them are to do with magnetic resonance imaging changes. Bone contusion is not a currently used notion. However, it was mentioned in great detail by Hippocrates and he stated that it could not be detected with the naked eye but was assumed to be present around any fracture. Apparently, in any patient needed black dye to determine the presence or absence of a fissure, the bone around the fissure would take up the dye and this Hippocrates considered was contused bone.

The term fell into desuetude until Le Dran who used it in patients who suffered infection and Pott who did the same. There is no common etiological feature in Le Dran's 3 and Pott's 12 cases. All cases suffered intracranial extra cerebral infections. The pathological nature of these contusions remains unclear as the sort of infections Pott and Le Dran saw no longer occur.

VASCULAR ANATOMY

It is well known that Galen described a vascular supply to the brain which included a rete mirabilis instead of a Circle of Willis. This error was corrected by Vesalius. In fact, yet again Berengario Da Carpi was ahead of everyone else since he wrote that he could find no such rete long before Vesalius published his works. Even so, as before, his observations did not become part of standard medical knowledge. In the present context, this is of little importance. However, there is one curiosity upon which there is little or no comment. Galen worked on many animals including Macaque monkeys. These monkeys have a circle of Willis.[1] Yet Galen stuck with his rete.

CEREBRAL ANATOMY AND PHYSIOLOGY

There was no means for proper study of cerebral anatomy before the invention of fixatives, stains, and microscopes. The regions of the brain were known including the separation into cerebrum and cerebellum. There were a number of discreet areas of understanding or perhaps better to say sets of concepts. They did not summate into a cohesive whole. They did not really interrelate at all. This makes the history of brain structure and function confusing. Those sets of concepts which are relevant to this text are outlined later.

Hippocrates knew that the brain was the organ of the intellect and the emotions. Galen regarded it as the seat of the soul which he thought was in the parenchyma but about the precise nature of the relationship between soul and brain he professed ignorance. Intellectual function was divided into Sensation, Intellect, and Memory. Galen moreover considered there were three essences which he called pneumata, one of which was manufactured and present in the brain. It was called animal or psychic pneuma and was the means by which the body interacted with the soul. He showed that various experiments in which the brain was reversibly damaged led to loss of consciousness followed by reawakening. This he considered was mediated by the psychic pneuma. However, it does not seem to have occurred to him that loss of consciousness after cranial trauma was due to effects on the brain. Thus, his management of cranial trauma involved observation, surgery, and bloodletting which had its origins in the physiology of the four humors as devised by Hippocrates. Their amounts had to be kept in balance and letting out blood should help to restore the balance. In the event of a minor disturbance, the bloodletting could be replaced by purgation. For hundreds of years, these measures where applied and their lack of effect had no effect on the application of accepted wisdom.

The work of Celsus was lost but when it was rediscovered at the end of the 15th century. While there was much excellence in this text it was not helpful in respect of cranial trauma since Celsus taught that what we now know to be symptoms of cerebral injury were due to damage to the cranium and meninges. This inaccurate notion was widely accepted.

The early church fathers edited Galen's notions of cerebral function. They devised a teaching called "Cell Doctrine" which moved the components of mental function from the parenchyma to the ventricles. The lateral ventricles had sensation at the front and imagination at the back. Intellect was in the third ventricle and memory in the fourth. Variations in this system were not uncommon.

BRAIN IDENTIFIED AS SOURCE OF CEREBRAL SYMPTOMS

While Hippocrates thought that damage to the pia and the cerebrum within was inevitably lethal, Galen disagreed and provided evidence for the correctness of his view. Centuries later, this was expanded. Paré recorded that parts of the brain could be lost and yet the patient could continue to function normally. Wiseman observed the same thing and reached the same conclusion. Thus, there was a body of thought for believing the brain could not be the source of symptoms following trauma. It was finally the French school which identified the brain as the source of the symptoms using strict logic. It was argued that the symptoms could not be due to skull fractures or meningeal tears the symptoms improved while the fractures and tears remained unchanged. From then on it was fairly quickly understood beyond doubt that alterations in consciousness, headache, vomiting, and paralyses were all a result of damage to the brain.

CEREBRAL LATERALIZATION

It is a commonplace today that damage to one side of the cerebrum produces paralysis on the opposite side of the body. However, there is no evidence that Hippocrates, Celsus, and Galen were aware of this relationship. Hippocrates did notice injury on one side could produce epilepsy on the opposite side. This contralateral relationship between injury and loss of function is so striking, it is strange that it was not noticed by any surgeon with wide experience of cranial trauma but such is not the case. In classical times, there was one exception, Aretaeus the Cappadocian. He was aware of the relationship and explained it by a crossing of nerves in an X. His work for whatever reason was ignored.

In the succeeding centuries, William of Saliceto in the 13th century observed that injury on one side was associated with paresis on the other. Over 200 years later Berengario Da Carpi quoted Avicenna as stating that paralyses and trauma occurred on the same side. He went on to state that in his experience a paralysis could be both ipsilateral or contralateral. Paré was aware of contralateral epilepsy but did not comment on paralysis. In the 18th century, Lorenz Heister (1683−1758) mentions it in plain text that paralysis after trauma was contralateral to the injury. James Hill was sufficiently aware of contralateral paresis that he tried to use this knowledge in his surgical planning. Benjamin Bell, Hill's one time student was aware of contralateral pareses. He also mentioned fixed dilated pupils which was a first. Abernethy specifically mentions that injury on one side produced paralysis on the opposite side. However, it was Sir Jonathan Hutchinson who finally provided evidence which gained acceptance for this relationship in 1857. In this context, it is worth noting that the matter of contralateral paralysis was determined a full 20 years before the debate about cerebral localization was finally decided.

CONCLUSION

The contents of this chapter thus outline a variety of notions which were relevant for the surgical management of cranial trauma and how they developed over the centuries. One other phenomenon remained unexplained and that was the lucid interval between trauma and clinical deterioration. This is the topic of the next chapter. Its history has also been confused[2] but it has great importance as it was a search for the underlying mechanism which formed the basis of the experimental studies which form the final section of this book.

REFERENCES

1. Kumar N, Lee JJ, Perlmutter JS, Derdeyn CP. Cervical carotid and circle of Willis arterial anatomy of macaque monkeys: a comparative anatomy study. *Anat Rec (Hoboken)*. 2009;292 (7):976−984.

2. Ganz JC. The lucid interval associated with epidural bleeding: evolving understanding. *J Neurosurg*. 2013;118(4):739−745.

CHAPTER 11

The Lucid Interval

THE REALIZATION OF THE INTERVAL

It was Lorenz Heister (1683–1758) who first suggested that symptoms due to cerebral compression could occur after a delay. The surgeons of the Académie Royale de Chirurgie, led by Jean Louis Petit, expanded the importance of timing in the development of symptoms after head trauma. Petit's colleague Henri-François Le Dran (1685–1770) was the first to publish these ideas in English in 1740 since Petit's papers were not published until long after his death. The group as mentioned earlier demonstrated that changes in consciousness and paralyses were due to lesions of the brain. In addition, they distinguished between symptoms arising from concussion (which were immediate) and those arising from compression (which could be delayed). They also realized the two groups of symptoms could merge into each other.

LE DRAN AND POTT

In Le Dran's series of 14 patients, there were three with epidural hematomas (EDHs) of whom none had a lucid interval. Percival Pott had three patients in his material with EDH none of whom had a lucid interval. On the other hand, there were 13 patients with a lucid interval none of whom had an EDH but all had an infection.[1] It should be mentioned that the duration of the interval varied between 10 and 29 days. This is not the same as the lucid interval following intracranial hemorrhage which is usually several hours up to a day or two. There are exceptions but an interval of more than 5 days is very unusual.

FIRST REAL CASE WITH A LUCID INTERVAL

The first real case published with a lucid interval associated with an intracranial hemorrhage was Case IV of James Hill.[2] The interval was 3 days and there was an intracranial hemorrhage only it was not an

Intracranial Epidural Bleeding. DOI: https://doi.org/10.1016/B978-0-12-812159-7.00011-4

EDH. It was a subdural hematoma. The first recorded lucid interval due to an EDH was Case XII of John Abernethy.[3]

ACCEPTANCE OF THE LUCID INTERVAL

The lucid interval was clearly described in the John Erichsen's surgery text book in the 1850s,[4] indicating it was widely accepted by the middle of the 19th century. However, the final definitive statement about the lucid interval was made by Sir Jonathan Hutchinson in his lectures on compression of the brain. He stated "It is certain that the value of the lucid interval before coma, as a symptom of ruptured meningeal artery, can scarcely be overrated. It is worth all the rest together."[5]

REFERENCES

1. Ganz JC. The lucid interval associated with epidural bleeding: evolving understanding. *J Neurosurg.* 2013;118(4):739−745.

2. Hill J. *Cases in Surgery.* Edinburgh: J Balfour; 1772.

3. Abernethy J. *Surgical Observations on Injuries of the Head.* London: Longman; 1810.

4. Erichsen J. *The Science and Art of Surgery.* 2nd ed London: Walton and Maberly; 1857.

5. Hutchinson J. Four lectures on compression of the brain. *Clin Lect Reps London Hosp.* 1867;4:10−55.

SECTION *IV*

Pathophysiology

CHAPTER 12

Developing Notions of Pathophysiology

THE PROBLEMS

In this chapter attention is focused on the isolated epidural hemorrhage. Of course, the clinical picture can be compounded by the presence of other brain injuries and confused following the intake of alcohol. These aspects of the condition are ignored in the current chapter. The details of the injuries in any given patient are no longer a source of uncertainty, since a computed tomography examination will reveal all the acute intracranial injuries both in and outside the brain. Epidural hemorrhage is different from other intracranial hemorrhages in two major respects. First, the bleed occurs into what is called a potential space. There is nowhere for an epidural hematoma (EDH) to exist before dura is separated from its normal location attached to the inner surface of the skull. The second characteristic of these bleeds is the lucid interval. The development of understanding of this phenomenon is outlined in an earlier paper[1] and repeated in previous chapters. It is known that lucid intervals occur with other intracranial bleeds. It is also known that they are not present in every case of EDH. The important point is that they are far more common with EDH than with any other intracranial bleed. The assumption has almost always been that secondary deterioration after an interval is the result of persistent extravasation from a torn artery with concomitant increase in hematoma volume.

Thus, studies of the pathophysiology of intracranial epidural bleeding need to examine the initial formation of a hematoma after a blow to the head. This will need to include studies of the mechanisms underlying the separation of the dura from the skull. It will also need to examine the mechanisms permitting ongoing intracranial hemorrhage. However, these concerns cannot be examined in a vacuum. They occur within the cranium and the mechanisms by which intracranial pressure is controlled and varied must also be considered. Finally, studies will not only concern the formation of the hematoma but also its effects on the brain and how these relate to survival.

Intracranial Epidural Bleeding. DOI: https://doi.org/10.1016/B978-0-12-812159-7.00012-6

EARLY DEBATE ON PATHOPHYSIOLOGY
OF EPIDURAL BLEEDING

There had been glimmers of pathophysiological understanding in the ancient world. Hippocrates wanted to let out collections of blood between the bone and dura in children. Celsus was aware that hematomas could develop without a fracture. However, the teachings of Galen followed by so many of his successors would have made the correct understanding of how EDHs impossible. The incorrect concept was the notion that the dura was attached to the skull only at the sutures and that the pericranium formed from the dura via the sutures and that its tense application to the outer surface of the skull held the dura in place. Giacomo Berengario Da Carpi (1460–1530) was the first to insist that the dura was evenly adherent to the interior of the skull, but his teaching was ignored. It was not until the early 18th century, when it was understood that the brain was the source of symptoms following injury that among others Percival Pott insisted that the dura was densely adherent to the interior of the skull.

The denseness of these adhesions resulted in a heated debate in the early 19th century. The celebrated participants in this debate were Sir Charles Bell (1774–1842) and John Abernethy (1764–1831). It will become clear that at the time of this debate, medical men disagreed in public with an intensity and candor which would never be permitted by a modern editor. Bell was the fourth son of an Episcopalian minister in Edinburgh. He claimed he received all his basic education from his mother. His surgical education he received largely from his elder brother John. However, George Bell, who was older than Charles but younger than John, advised Charles to travel to London, as John's quarrelsome nature had made him sufficiently unpopular that it could have affected Charles' career in Edinburgh. Thus, Charles became yet another Scot to have a profound effect on the development of medical science in the English capital. From what follows, it will become clear that while not as toxic as his brother John, he was nonetheless quite capable of being cantankerous himself.

It is necessary to start with Abernethy's writings to which Bell would react with such energy. Abernethy was himself of Scottish and Irish ancestry. He was initially trained by an incompetent surgeon called Sir Charles Blicks.[2] Thereafter, he was influenced by Sir William Blizzard at the London Hospital, to whom he dedicated the book in which his three cases of EDH are described. He was one of the distinguished London surgeons later influenced by yet another expatriate Scotsman: John Hunter. Abernethy described a series of 17 cases of head injury. Case VII was a man knocked down by the hook of a crane. He was stunned but then got up and walked home and put himself to bed. Later that same day a surgeon was called, as the man was unconscious. He was surgically treated, and a large EDH was found; after removal of the EDH, the brain did not reexpand and the patient died. The middle meningeal artery was injured but did not rebleed following removal of the hematoma. This is a description of a genuine lucid interval. Abernethy proposed that "it seems then that extravasation between the dura mater and the cranium is almost the only case which admits of being remedied by the use of the trephine." His comments on epidural bleeding include the following. "But if the fracture happens in the track of the principal artery of the dura mater, if the trunk or even a considerable branch of that vessel be torn, the hemorrhage will be profuse, and the operation of the trephine become immediately necessary to preserve the life of the patient."[3] He is stating that the hematoma is the result of hemorrhage from a torn middle meningeal artery or one of its major branches. There is no specific comment on delayed ongoing bleeding enlarging hematoma volume. The issue of the timing of the hemorrhage is not mentioned.

Bell would not accept Abernethy's implied contention that ongoing bleeding from a middle meningeal artery could lead to the formation of a hematoma. He begins by stating of Abernethy, "It will appear no doubt remarkable that he should display on many occasions such *vigour* of intellect and close reasoning, which have been attended with great improvements, and on this take so much for granted, and convey an advice so full of danger." He continues a little later, "I might say it is extraordinary that any one who had ever raised the skull-cap in dissection, and felt the strength of the universal adhesions of the dura mater to the lower surface of the bone, could for an instant believe that the arteria meningea media has power of throwing out its blood, to the effect of tearing up these adhesions from the entire half of the cranium!" Bell then recounts an elegant experiment in which he struck a head with a mallet. He then injected size (a fluid wax that sets once cooled). The wording he uses is obscure. He states "... inject the head minutely with size injection and you will find a 'clot' of the injection lying betwixt the skull and the dura ...."[4] There is no statement as to the root of injection. If the head were opened prior to direct injection this would mean one thing. This author has always assumed that such an injection was made up the carotid artery. If that is the case the leakage of size into the epidural space would require the rupture of arteries. The point remains obscure. He deduced not unreasonably that the blow has separated the dura from the bone and concluded that "the extravasation of blood is a consequence and not the cause of the separation." He concluded with the vital remark, "It proves also that the advice is wrong which would hurry the young surgeon to trepan the skull immediately, to preserve the life of the patient, and to prevent the hemorrhage from being profuse." It may be mentioned in conclusion that Bell's experiment also demonstrated how an EDH can develop without a fracture. On the other hand, he does not state an opinion on the timing of the hematoma. His argument relates to the initial generation of the hematoma and its ability to increase in volume due to ongoing arterial hemorrhage. His advice is also in keeping with his brother John's abhorrence of trepanation.

SIR JOHN ERICHSEN 1895

The debate between Bell and Abernethy highlights a factor of vital importance in the evolution of an EDH. It may be considered that at the time of impact a potential epidural space is opened which may fill with blood from a ruptured major meningeal artery. Subsequently, it is accepted today that further expansion of the hematoma may occur as a result of ongoing bleeding from the ruptured artery.

In the 10th edition of his massive textbook on surgery Erichsen again addresses the issue of epidural bleeding.[5] He repeated the description of the clinical picture of initial unconsciousness, latent interval followed by increasing loss of consciousness. Hemiparesis or general paralysis would develop with dilated pupils. However, he went further to consider the mechanisms of bleeding; the first to do so since Bell and Abernethy. He made assumptions about the pressure in the artery. He then assumed the area of dura separated. He suggested that the force acting on the dura is amplified according to the principles of the hydraulic press. This means that in a closed fluid system the pressure is constant throughout. If a force (F_1) is applied in one location over an area A_1 the force F_2 acting on area A_2 will be $F_1 \times A_2/A_1$ as shown in Fig. 12.1. This is ingenious but only of limited value because

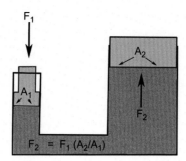

Figure 12.1 This illustrates how when a force is applied to an enclosed fluid, where the pressure is the same at all points the force applied at the large area A is amplified as indicated. This could be a means whereby bleeding from a small meningeal artery could affect ongoing dura separation.

the forces concerned were guessed at and not measured. Nonetheless, as will be shown below it was a step in the right direction. It was used by Erichsen to suggest that there would be enough force from the hemorrhage from a middle meningeal artery to continue to separate more dura from the skull, thus contradicting Bell's view. This conclusion is reached because Erichsen's approach is to state that there is no doubt that a meningeal artery can strip dura. He then presents some assumed numbers to support a clever argument. Thus, his approach remains more qualitative than experimental but despite its defects it was an important new attempt to understand the formation of EDHs. Its value was approaching the problem from a scientific direction. Finally, it may be mentioned that earlier in his account of these bleeds he mentioned that the source of the bleeding could be from the meningeal arteries or venous sinuses. This is speculative and as will be shown problematical. However, it brings us to the next contribution.

FREDERICK WOOD JONES AND VENOUS HEMORRHAGE

In 1912 Frederick Wood Jones (1879—1954), a senior anatomy demonstrator at St. Thomas' Hospital in London,[6] extended the paradigm. He noted that the grooves in the skull for meningeal arteries were in fact filled far more by veins than by arteries. He proposed that rupture of these veins could be the source of EDHs. He noted that local fractures of the skull inevitably led to damage to the veins but that the arteries in some instances could escape rupture. He examined three patients who had died as a result of such a bleed and the middle meningeal artery was intact in all of them. It will be subsequently shown that venous bleeding cannot expand an EDH; at least not in the closed skull with a normal intracranial pressure at the time of injury. Thus, Wood Jones paper needs careful reading to explain how it would seem that venous bleeding had produced a hematoma. There are a number of aspects to consider. It is feasible that in all three patients the initial impact had loosened the dura over a wide area. In the absence of any other factor, one might consider that venous blood could fill up the epidural space so created. There is however no evidence that venous bleeding could have led to ongoing extravasation with increasing hematoma volume and concomitant further stripping of dura from the inside of the skull. The paper's value lies in its demonstration that meningeal vascular damage may involve the veins as well as the arteries. However, it tells nothing about secondary extravasation with a venous origin. The notion of ongoing bleeding having a venous origin is a vexed one to which further attention will be addressed.

MR. PAUL'S HUNTERIAN LECTURE IN 1955

Major Milroy Aserappa Paul (1900−89) was a leading Sri Lankan surgeon who was the first Professor of Surgery at the Ceylon Medical College and a cofounder of the International College of Surgeons. His father was a surgeon and his mother the daughter of a physician. In 1955 his third Hunterian oration had for its subject "Hemorrhages from Head Injuries."[7] Paul opened his consideration of EDHs with two questions. Why does an untreated epidural hemorrhage continue bleeding till death supervenes? Why does simple evacuation of the hematoma arrest such hemorrhage? It is of some interest that neither of these questions examines a real phenomenon. Epidural hemorrhage may persist or cease. If they cease spontaneously the patient may well survive. Simple evacuation of the hematoma by no means arrests the hemorrhage. Indeed, it is a commonplace that hemostasis of the damaged artery is often the trickiest part of the operation. However, Paul argues a special case as shown later.

Initial Dura Separation

Paul repeated the Bell experiments at the vertex and at the side of the skull. Bell had found that if the blow of a mallet was at the vertex or occipital region a linear fracture resulted. On the other hand, if it were on the temporal bone there would be fracture with depression.[4] Paul found that using a heavy wooden plank, mallet or small metal hammer

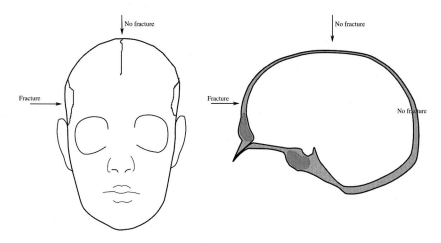

Figure 12.2 Blows to the vertex produced no fracture or dura separation. By contrast, blows to the frontal and temporal region do produce fractures and dura separation. There was however a fundamental difference in the setup. With vertex blows, the cadaver was supported in the sitting position. With the other blows, the head was lying respectively on occipital or contralateral temporal support. This means the side opposite the trauma has rigid support This situation would never arise in natural trauma. It would change the way the skull shape was changed by a blow.

failed to produce a fracture or separation of dura at the vertex. In the other hand a blow to the midline frontal region and temporal regions produced fractures with dura separation. There was however a major difference when the vertex compared with the temporal and frontal regions as illustrated in Fig. 12.2. Since Bell did produce linear fractures when the vertex was struck maybe the head was supported in a different way or maybe he simply struck it more forcibly than Paul did.

Progressive Dura Separation

The injury following a frontal blow produced a dura separation over an area with a diameter of about 3 in. This is a rather limited area of separation. This localized separation is insufficient to account for the size of hematomas found at surgery. Paul reasoned from this that further dura separation must take place. He applied 200 mm Hg static pressure for through trephine openings of two different sizes. The smaller one with a half inch diameter and the larger with a 2-in. diameter. There was no dura separation through the smaller hole but substantial separation with the larger. Paul then used finger dissection to assess the force needed to strip the dura off the skull and commented that the force needed is not great; except close to the longitudinal sinus and medial portions of the skull base. He correctly concluded that the area where dura separation by the finger is easiest closely follows the distribution of clinical EDHs. There is a thorough account of the clinical picture in Paul's manuscript which contains nothing new of importance.

Effect of EDH on the Intracranial Pressure

Paul devoted one and a half pages to speculation about the effect of the hemorrhage on intracranial pressure. The arguments are ingenious but supported by no evidence of any kind. Later work has indicated that there is much error, so there is no reason to continue to recount the opinions expressed in this part of the manuscript.

Arrest of Epidural Hemorrhage

Paul made the following two remarkable statements. "If evacuation of a blood clot is followed by a torrential hemorrhage, the hemorrhage will be either from the superior longitudinal sinus or from the lateral sinus." This is followed by "In a consecutive series of 33 cases of extra dural hemorrhage under my care the bleeding had occurred either from the meningeal veins or from the large venous sinuses. Hemorrhage from other sources has never been encountered by me." This is so at odds with conventional experience there is no easy explanation.

Assessment of Paul's Work on Epidural Bleeding

While Paul's paper was extensive there are portions which are rather bizarre. He stated that the bleeding pressure is below the venous blood pressure without a scrap of evidence. He stated that hydrostatic pressure cannot be transmitted through a blood clot in the manner presented by Erichsen because it is almost dry which to put it mildly is odd. He suggested without evidence that epidural bleeding from low pressure sources would continue until death unless the hematoma is removed. His comments on sources of bleeding and hemostasis do not match common experience. Nonetheless, his paper is a major attempt at illumination of pathophysiology. It has however remained obscure because it is sad but true that the Annals of the Royal College of Surgeons of England are perhaps not that widely read by neurosurgeons.

THE FIRST SCIENTIFIC ASSESSMENT OF TRAUMATIC SEPARATION OF THE DURA FROM THE SKULL

The opportunity is taken to repeat what is the general experience. Most temporal EDHs are associated with a ruptured middle meningeal artery which most often will require hemostasis during surgery. This takes a bit more time when the artery passes through a channel in the bone. The multiple points of bleeding from the dura, revealed when the hematoma is removed, would seem to be mainly from tiny arteries since the blood is invariably oxygenated. They are mostly controlled with the application of saline compresses and in a few cases with the use of bipolar diathermy. However, it is emphasized that the common experience when operating on posttraumatic temporal EDHs is that the bleeding is arterial.

SCIENTIFIC PATHOPHYSIOLOGY—FORD AND MCLAURIN

The first genuinely scientific analysis of the pathophysiology of epidural bleeding was published in 1974 by Ford and McLaurin.[8] This paper examined the physical requirements which would enable separation of dura from the skull. It was not concerned with initial dura separation mechanisms. The experiments were carried out in anaesthetized dogs.

Burr holes of varying sizes were drilled and a #18 short needle was placed with its tip in the epidural space and it was the fixed by means of acrylic used to seal the burr hole opening. This is illustrated in Fig. 12.3. Tubing attached to the needle permitted the application and

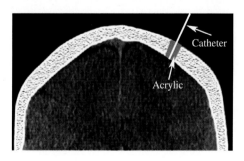

Figure 12.3 The burr hole opening through the dog skull filled with acrylic through which passes a catheter via which force can be applied to the dura and via which the epidural pressure may be measured.

measurement of pressure. The pressure in one series of experiments was applied from a steady source. In the other series, it was pulsatile with the femoral artery being used as the source. The force applied to the dura would vary with the amount of pressure applied and the area of dura dissected free. They found that for a steadily applied force to produce reliable separation of dura a force of 90 g was required and for a pulsatile force the value was 35 g. The relevant equation is:

Force applied to dura in grams = input pressure $(g/cm^2) \times$ area separated (cm^2).

This means that for a given input pressure, the force applied will be determined by the area over which it is applied, that is to say the area of dura already separated from the skull. It became clear to the authors that an area of dural separation with a diameter between 6 and 8 mm was enough to permit the arterial pulse to separate more dura. With a normal intradural pressure no dura separation could be effected with the kind of pressures generated by venous bleeding. It should be noted these statements are made without documentation. It is just something stated, but the authors take it seriously and include it in the conclusions of the paper.

DURA ATTACHMENT MEASURE IN MOSCOW

There is one other study of relevance. This is from the Burdenko Institute in Moscow.[9] This study quantified the range of the strength of fixation of the dura to the skull in the corpses of young children and mature adults. The range was enormous varying between 30 and 1800 gm/cm. This variability could go some way to explaining the variable clinical course of epidural bleeding.

REDUCED INTRADURAL PRESSURE

There is one special situation where epidural bleeding occurs without trauma and arterial injury. That is when there is a substantial and sudden drainage of ventricular cerebrospinal fluid (CSF). This was reported early in a study of air ventriculography where substantial volumes of CSF were extracted prior to the introduction of air. Frontal EDHs developed.[10] There have also been a number of reports of EDHs occurring following shunting for hydrocephalus.[11-14] It is the

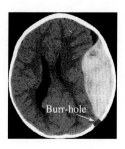

Figure 12.4 Characteristic location of an epidural hematoma complication ventricular puncture during a shunt operation.

author's experience that this unusual complication of shunting occurs with a characteristic anatomical localization as shown in Fig. 12.4.

It is suggested the reason that EDHs can occur following ventricle drainage is due to a combination of reduced intradural pressure due to fluid removal and weak adhesion between the dura and the skull in that particular patient. While the pressure gradient from outside to inside the dura will not be all that great, it is applied over the entire area of one side of the brain. This means from the earlier equation that the force across the dura could be considerable. Necessary measurements to confirm this notion have never been made so that the suggested mechanism is speculative. The suggestion is however logical within our current understanding of the formation of EDHs expressed in this and subsequent chapters.

CONCLUSION

Thus, by the end of the 1970s ideas had reached the following stage. EDHs are mostly caused by an arterial rupture into an epidural space created as a result of a blow to the head. The size of the hematoma would depend on the amount of dural separation and the intradural pressure resisting further separation which in turn would depend on concomitant subdural/intracerebral injury. Thus, much had been understood about dural separation. However, nothing had been done to explain the mysterious latent interval followed by delayed deterioration and often death. Studies relating to this phenomenon require an understanding of normal intracranial vascular dynamics and the reactions to volume loading with in the cranial cavity and these matters will be examined in the next chapter.

REFERENCES

1. Ganz JC. The lucid interval associated with epidural bleeding: evolving understanding. *J Neurosurg.* 2013;118(4):739−745.

2. Jacyna L. *Abernethy, John (1764−1831).* Oxford: Oxford University Press; 2004. Available from: http://www.oxforddnb.com/view/article/49.

3. Abernethy J. *Surgical Observations on Injuries of the Head.* London: Longman; 1810.

4. Bell C. *Surgical Observations.* London: Longmans; 1816.

5. Erichsen J. *The Science and Art of Surgery.* 10th ed. London: Longman and Green; 1895.

6. Wood Jones F. The vascular lesion in some cases of middle meningeal haemorrhage. *Lancet.* 1912; July 6:7−12.

7. Paul M. Haemorrhages from head injuries. *Ann R Coll Surg Engl.* 1955;17(2):69−101.

8. Ford LE, McLaurin RL. Mechanisms of extradural hematomas. *J Neurosurg.* 1963;20:760−769.

9. Murzin VE, Goriunov VN. Study of the strength of the adherence of the dura mater to the bones of the skull. *Zh Vopr Neirokhir Im N N Burdenko.* 1979;4:43−47.

10. Ameli NO, Sodeify N. Anterior fossa extradural haematoma following ventriculography through posterior Burr-holes. *Acta Neurochir (Wien).* 1965;13(3):464−468.

11. Alsheheri MA, Binitie OP. Acute epidural hematoma following restoration of ventriculoperitoneal shunt patency. *Neurosciences (Riyadh).* 2004;9(4):312−314.

12. Kalia KK, Swift DM, Pang D. Multiple epidural hematomas following ventriculoperitoneal shunt. *Pediatr Neurosurg.* 1993;19(2):78−80.

13. Noleto G, Neville IS, Tavares WM, Saad F, Pinto FC, Teixeira MJ, et al. Giant acute epidural hematoma after ventriculoperitoneal shunt: a case report and literature review. *Int. J. Clin. Exp. Med.* 2014;7(8):2355−2359.

14. Tjan TG, Aarts NJ. Bifrontal epidural haematoma after shunt operation and posterior fossa exploration: report of a case with survival. *Neuroradiology.* 1980;19(1):51−53.

CHAPTER 13

Intracranial Vascular Dynamics

INTRACRANIAL PRESSURE

The most important item related to the pressure inside the head is the maintenance of cerebral blood flow (CBF). This is normally around 55 mL/100 g tissue/min; one of the highest flows in the body. Its maintenance must take into account that the brain is enclosed in a rigid box, the cranium. Very roughly the brain takes up 87% of the intracranial contents, the cerebrospinal fluid (CSF) accounts for 9% and the blood, at any one time mostly in the venous side of the circulation accounts for 4%.

In his book "Observations on the Structure and Function of the Nervous System," Alexander Monro Secundus (1733−1817) wrote as follows. "For as the substance of the brain, like that of the other solids of our body, is nearly incompressible, the quantity of blood within the head must be the same, or very nearly the same, at all times, whether in health or disease, in life or after death, those cases only excepted, in which water or other matter is effused or secreted from the blood-vessels, for in these, a quantity of blood, equal in bulk to the effused matter, will be pressed out of the cranium." A while later George Kellie (1720−79) provided experimental confirmation in animals, showing that following death from cyanide, draining blood from the animal produced exsanguination of extracranial tissues but the brain retained its blood. Thus, this notion that the change in the volume of one intracranial component could only be achieved through a complementary change in one of the other components came to be known as the Monro-Kellie Doctrine. The normal mean intracranial pressure (ICP) in a recumbent patient measured following puncture of the CSF spaces in the lumbar region is maintained at between 7 and 15 mm Hg. It is maintained by the blood pressure.

Intracranial Epidural Bleeding. DOI: https://doi.org/10.1016/B978-0-12-812159-7.00013-8

CEREBROSPINAL FLUID

Within the head is a crystal-clear colorless fluid manufactured in the ventricles and called CSF. There are descriptions of water in the head in the writings of the ancients. Thus Hippocrates mentioned "water over the head."[1] Celsus mentions "... there is a class which may become chronic, in which a humour inflates the scalp so that it swells up and yields to the pressure of the fingers. The Greeks call it hydrocephalus."[2] Galen incised the scalp for water under the skin. He performed a trepanation for water under the cranium. Water within the dura was untreatable.[3] He considered only the presence of such water in infants and considered it the result of inappropriate squeezing by midwives.[1] Interestingly, nobody paid much attention to this fluid for centuries. This may in part be due to mistaking it for a humor as indicated by Celsus. One author speculated that it might be the result of autopsy technique, where separating the head from the body could result in loss of blood and CSF.[4] Be that as it may, real appreciation of the presence of CSF came with the work of Emanuel Swedenborg (1688–1772) of Uppsala in Sweden. To understand the structure and function of the CSF, it is necessary to understand its secretion, circulation, and absorption.

Secretion of the Cerebrospinal Fluid

Vesalius (1514–64) noted that in hydrocephalus, the water accumulated not under the skull but in the ventricles. Thomas Willis (1621–75) first suggested that CSF was produced by the choroid plexuses.[5] François Magendie (1783–1855) in 1825 described CSF adequately for the first time.[6] He established the connection between the ventricles and the subarachnoid space via the foramen in the fourth ventricle which bears his name. He established the correct direction of flow,[7] even though he incorrectly thought that the fluid was made by the pia.[4] Research of Faivre in 1853 and Luschka in 1855 suggested that CSF was indeed produced by the choroid plexuses within the cerebral ventricles. Pharmacological evidence was described in the first decade of the 20th century supporting the choroid plexus as a source of CSF. A few years later Harvey Cushing and Walter Dandy added considerable extra evidence to support what has become the accepted view.[6] The pattern of CSF flow is shown in Fig. 13.1.

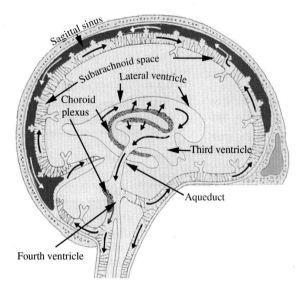

Figure 13.1 This diagram illustrates the essential anatomy of CSF production, circulation, and absorption. From The History of the Gamma Knife, with permission from Elsevier.[8] See text.

Modern teaching is that CSF is produced mainly by the choroid plexus in a two-step process. The first step is passive filtration of plasma from the choroidal interstitial compartment. The second step involves active transport from this compartment to the lumen of the ventricle. There is some extra-choroidal secretion. Some is derived from the extracellular fluid and brain capillaries across the blood brain barrier. It is of little physiological significance. Some CSF may also be produced by the ependymal epithelium. This secretion can be altered by changes in the ependyma especially if the ventricles are dilated.[9]

Absorption of the Cerebrospinal Fluid

This was the subject of early debate. It is assumed that production of CSF was constant.[6] So, there must be also constant absorption to maintain a steady ICP. Antonio Pacchioni (1665–1726) had described the granulations which bear his name in 1705. In 1875, Axel Key (1832–1901) and Magnus Gustaf Retzius (1842–1919) published a beautiful illustrated monograph about the brain and its membranes. They believed that the Pacchionian granulations were the location of CSF absorption. In 1922, Lewis H. Weed (1886–1952) published his alternative absorption mechanism with its elegant experimental basis.

Weed and Cushing distinguished between larger relatively sparse Pacchionian granulations and multiple microscopic arachnoid villi.[6,10] Weed demonstrated that beyond reasonable doubt, the absorption was through the multiple tiny arachnoid villi.

Modern teaching is little changed from the time of Weed. However, the arachnoid villi are not just located around the sagittal sinus but also along the spinal nerve roots. CSF can also be absorbed by cranial nerve sheaths, the ependyma, and extracellular fluid according to pressure gradients. Part of the CSF is absorbed by the olfactory mucosa and cranial nerve (optic, trigeminal, facial, and vestibulocochlear nerves) sheaths and is drained by the lymphatic system. At normal ICP, 10% of cervical lymph is derived from CSF, but when ICP increases from 10 to 70 cm H_2O, 80% of cervical lymph is derived from CSF. The function of this lymphatic pathway is unknown.[9]

The Circulation of the Cerebrospinal Fluid
The term circulation of the CSF was introduced by Harvey Cushing in 1925.[10] The reasoning is straightforward. On the one hand, there is constant production of CSF with a stable ICP. On the other hand, experiments by Cushing and Walter Dandy (1886–1946) showed that blocking different narrow parts of the ventricles produced dilatation of the ventricles upstream.[6]

Cerebrospinal Fluid Volume Production and Resting Pressure
The earlier mechanisms are now accepted. CSF pressure is the result of their coordinated activities. The next question is what are the relevant quantitative values. The CSF is produced by the choroid plexus at an amazingly constant rate of around 0.35 cm^3 per min. The overall volume of adult CSF is around 150 cm^3 of which around 25 cm^3 is in the cerebral ventricles. The secretion rate mentioned earlier produces roughly 600 cm^3 a day so that the total CSF volume is produced four times a day. The resting CSF pressure in adults is 10–15 mm Hg.

CEREBRAL BLOOD FLOW
Basic Cerebral Blood Flow
The brain has a high blood flow of around 55 mL/100 g/min. The gray matter flow is higher at 75–80 mL/100 g/min and white matter flow is less at 20–30 mL/100 g/min. The pressure driving the blood round

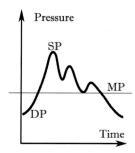

Figure 13.2 The difference between the systolic and diastolic pressures.
DP, *diastolic pressure;* SP, *systolic pressure;* MP, *mean pressure (DP + 1/3(SP − DP))*
The mean pressure is calculated in this way since studies show that diastole takes up most of the time during a single heart beat and the pressure is gradually falling and this way of registering mean pressure is close to the reality.

inside the cranium is not constant but pulsatile reflecting the pulsatile activity of the heart. In consequence, it is helpful to have a pressure value which is independent of varying pulses. This is the mean pressure which is the diastolic pressure plus 1/3 of the pulse pressure (the difference between the systolic and diastolic pressures, see Fig. 13.2 for explanation). The pulse pressure of the CSF has a similar shape. The pressure driving blood round the brain is the blood pressure minus the ICP. This is the cerebral perfusion pressure (CPP) calculated from the simple equation as shown in the next line.

$$CPP = Mean\ BP - Mean\ ICP$$

To maintain and adequate CBF, the CPP must be maintained ideally over 60 mm Hg. According to Ohm's law, CBF is proportional to the perfusion pressure expressed CBF $\propto$ CPP. The flow of a fluid through a pipe or tube is determined by Poiseulle's law as shown in the next line.

$$Q = \Delta P \pi r^4 / 8l\eta$$

where Q is flow, ΔP is the pressure gradient, r is the radius, l is the length of the tube, and η is the fluid viscosity (in this case, the blood viscosity).

Autoregulation

It follows that the CBF is proportional to the driving pressure and the fourth power of the radius and is inversely proportional to the length of the tube and the viscosity of the blood. These may be considered

the anatomical or physical factors affecting CBF. In addition, it is reg-
ulated via two main physiological mechanisms, autoregulation and
arterial CO_2 tension. These mechanisms work by varying the diameter
of the arterioles of the brain which in view of Poiseulle's law with its
radius to the fourth power has a profound effect on flow. The characteristics of autoregulation are well known though the underlying
mechanisms by which it is implemented remain somewhat obscure.

The basic principle is that between certain limits changes in the
arterial pressure and hence the CPP are compensated and the CBF
remains stable and unchanging. This is illustrated in Figs. 13.3 and
13.4. They illustrate the relationship between CPP and CBF over a
given range of arterial pressure. They also show that lowering the
blood pressure in hypertensive patients is dangerous because it will be
easier to fall outside the functioning range of autoregulation and below
that lower limit, the CBF will fall following the CPP passively.

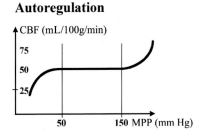

*Figure 13.3 CBF remains unchanged over a range of perfusion pressures by means of a process called
autoregulation.*

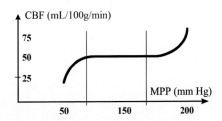

*Figure 13.4 CBF remains unchanged over a range of perfusion pressures. However, in the presence of hyperten-
sion, the tolerance limits are moved to the right.*

CBF – CO$_2$ response

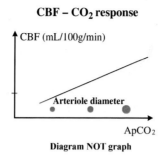

Diagram NOT graph

Figure 13.5 Changes of CBF in response to increasing carbon dioxide in the blood.

Cerebral Blood Flow and CO$_2$

Increasing the ApCO$_2$ causes vasodilatation as shown in Fig. 13.5. This is relevant in surgery and for a while it was customary to hyperventilate patients during craniotomy down to 50% of a normal ApCO$_2$. It worked because the contracted arteries and reduced CBF meant that there was less blood in the cranium and thereby in accordance with the Monro-Kellie doctrine, the ICP could be reduced and the surgical conditions made easier. However, when this marked hyperventilation was reversed, a reactive hyperemia occurred when the CO$_2$ was returned to normal. It could take a very long time before hemostasis could be achieved. Today, it is customary to ventilate the patient with just a slight fall in ApCO$_2$. It is also important in cases with multiple trauma and is one of the reasons why the first treatment in such cases is to establish the airway and ensure CO$_2$ is not allowed to rise with a concomitant rise in ICP and fall in CPP with a concomitant reduction in CBF.

CEREBRAL ISCHEMIA

For normal cerebral function to be maintained, the CBF should be kept as high as possible. Nonetheless, if the regulatory mechanisms fail and the CBF starts to fall, it must first fall to around 20 mL/100 g/min before neurons stop functioning. At this reduced level of blood flow, the cells lose their electrical function but stay alive. If there is a further fall to around 10 mL/100 g/min, then neurons start to die as evidenced by leakage of intracellular potassium into the extracellular fluid.

ACUTE INTRACRANIAL VOLUME ENCROACHMENTS

Having outlined the mechanisms whereby CBF is controlled, it is now necessary to proceed to situations where the control mechanism may be brought into play. The Monro-Kellie doctrine highlights the limitations governing the physiological responses to increasing the volume of any intracranial component. To a limited extent, blood and/or CSF can be displaced but when the limit of this displacement is reached, the pressure inside the head will increase and eventually CBF may be compromised. In the case of slowly increasing volumes, such as occur with the growth of a meningioma, there is a possibility for displacement of the brain which for a long time can compensate for the slowly growing tumor. Such a relief mechanism is not available for acute volume loading, which is relevant in the present context of epidural hemorrhage. Hemorrhage is the most rapid form of intracranial volume loading. However, over a slightly longer period, an established hematoma, dependent on its location may lead to extra volume loading due to the abnormal accumulation of water in the cranium. This water may be in the brain parenchyma in the form of cerebral edema. On the other hand, some hematomas may be in a location which blocks the passage of CSF through the ventricles leading to the development to hydrocephalus.

This brings us to the dynamics of intracranial hemorrhage. It was first as investigated by Nicolas Zwetnow (1929–2016). The crucial paper was written by Zwetnow and Löfgren in 1972.[11] What they demonstrated was that intracranial bleeds had a limited time course as shown in Fig. 13.6. This is not a true graph but a graphic diagram of

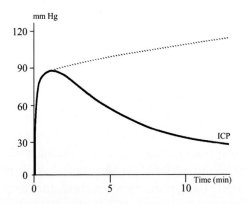

Figure 13.6 Changes in ICP following an intracranial bleed. The solid line shows what happens with normal blood and the dotted line shows what happens with heparinized blood. See text. NB: this is a diagram not a graph.

how intracranial arterial bleeding works. There is a different curve for venous bleeding but that is not relevant in the case of epidural hemorrhage so is not considered. It was found that the pressure from a significant arterial bleed rose very quickly and then after a couple of minutes starts to fall, so that the hemorrhage would be over within a few minutes. The reasons for this are as follows.

The pressure generating the hemorrhage, the bleeding pressure, is the arterial pressure (P_a) minus the ICP; whereas for epidural bleeds, it is the arterial pressure minus the epidural pressure (P_{epid}). As the intracranial compartment can only tolerate a limited volume loading, which is compensated for by pushing out blood and CSF as mentioned earlier, any further increase in an intracranial volume component will cause a rise in ICP. The volume component to be considered is the hematoma. As its volume rises, the ICP rises so that the bleeding pressure (Pa−ICP) falls and bleeding slows down to stop after about 8−10 min. If the blood is heparinized, the pressure continues to rise as there is no clot formation to stop the bleeding. Virchow's triad informs us that slow blood flow or stasis can facilitate the formation of intravascular blood clots or thrombosis. If the blood is not anticoagulated, then the rising ICP and concomitant falling bleeding pressure will facilitate the chances of coagulation with the formation of a clot which can stop the bleeding. Thus, the characteristic dynamics of an intracranial bleed is a sudden rise in pressure followed by a gradual fall over a few minutes. Following the cessation of bleeding, the pressure will remain higher for a period due to the presence of a hematoma mass. How much damage is done by a bleed will depend on its location and the final hematoma volume. Bleeds into deep parts of the brain can produce significant damage even with small volumes due to the concentration of function in the internal capsule and its neighborhood. Obviously, larger hematomas have a bigger chance of destroying nerve cells compared with smaller ones. Moreover, greater volumes of persisting hematoma mass can produce damage via ischemia following persistent raised ICP.

CONCLUSION

This book is concerned with epidural bleeding. Since this is superficial, its major effect on brain function will in the first instance be due to its volume raising ICP and putting the brain at risk for ischemic damage.

This mechanism of injury is complemented with larger hematomas by the incompressible hematoma distorting the incompressible but plastic cerebral tissue which herniates under the falx and through the tentorium producing increased damage and swelling in the herniated portions of the brain.

In the next chapter, it will be seen how epidural bleeds differ from the pattern described in this chapter because of the special anatomical characteristics of their location.

REFERENCES

1. Missori P, Paolini S, Curra A. From congenital to idiopathic adult hydrocephalus: a historical research. *Brain*. 2010;133:1836−1849.

2. Celsus. *De Medicina*. Cambridge, MA: Loeb Library, Harvard University Press; 1938.

3. Adams F. *Seven books of Paulus Ægineta*. London: The Sydenham Society; 1846.

4. Hajdu SI. Discovery of cerebrospinal fluid. *Ann Clin Lab Sci*. 2003;33(3):334−336.

5. Lifshutz JL, Johnson WD. History of hydrocephalus and its treatments. *Neurosurg Focus*. 2001;11(2):1−4.

6. Weed LH. The cerebrospinal fluid. *Physiol Rev*. 1922;2:171−203.

7. Herbowski L. The maze of cerebrospinal fluid discovery. *Anat Res Int*. 2013;2013:1−8.

8. Ganz JC. *The History of the Gamma Knife*. Amsterdam, London, New York: Elsevier; 2014:4−5.

9. Sakka L, Coll G, Chazal J. Anatomy and physiology of cerebrospinal fluid. *Eur Ann Otorhinolaryngol Head Neck Dis*. 2011;128(6):309−316.

10. Cushing H. The third circulation and its channels. *Lancet*. 1925;206(2):851−857.

11. Löfgren J, Zwetnow NN. Kinetics of arterial and venous hemorrhage in the skull cavity. In: Brock M, Dietz H, eds. *Intracranial Pressure*. Vienna: Springer-Verlag; 1972:156−159.

CHAPTER 14

Factors Affecting the Formation of Epidural Hematomas

INTRODUCTION

Throughout earlier chapters and particularly in Chapter 12, Developing Notions of Pathophysiology, mention has been made of the attachment of the dura to the skull and its relevance for epidural bleeding. There has however remained debate about how much of an epidural hematoma (EDH) arises immediately after trauma and how much develops later. Disagreement about this goes back to the squabble between Bell and Abernethy recounted in Chapter 12, Developing Notions of Pathophysiology.[1,2] The accounts of the subsequent history of notions related to the pathophysiology of epidural bleeding have regrettably been less careful than one might wish as will be related later. Up to the early 19th century, there had been no serious attempt to examine this topic. The Bell Abernethy debate was really the start. To recount, according to Bell, the dura separation occurring at the time of the injury was the sole factor which determined the volume of a hematoma. According to him, there was no possibility of secondary volume expansion from ongoing bleeding.

Repeated Errors in the 1960s

There is a reason to outline the relevant errors as they have left their mark on subsequent understanding of the processes whereby EDHs are formed. There is a reference in a much-quoted paper, from the *Journal of Neurosurgery* published in 1968 stating that Bell's experiments were preceded by a statement by Sir John Erichsen from 1779.[3] This is the first error since Erichsen was born in 1818 and the reference in question bears the date 1878.[4] It is really strange error as the as the relevant passage in Erichsen's paper states "Sir Charles Bell conclusively proved, by the experiment to which I have referred, that separation of the dura mater was the primary condition; and there can I think, be little doubt that the detachment of the dura mater is a result

Intracranial Epidural Bleeding. DOI: https://doi.org/10.1016/B978-0-12-812159-7.00014-X

of the blow on the head and the filling up is the consequence of that detachment and could not take place if the detachment had not previously occurred." This error was subsequently copied and repeated in another authoritative publication.[5]

The authors of the paper in which the abovementioned error occurred state that they repeated the Bell experiments with the same results. However, while Bell denied the possibility of ongoing bleeding causing further dura stripping the paper goes on to state the following without a shred of evidence "Once the hemorrhage has begun, the gathering clot fills the extradural pocket first and as the bleeding progresses the dura becomes stripped away in an ever-widening perimeter. In this manner, the hematoma can grow to enormous proportions extending all the way from the frontal to the occipital regions."[3] It is odd that the authors should on the one hand repeat Bell's experiments and then state that his conclusions were incorrect without providing evidence for this point of view.

It is also unexplained why this paper of 1968 did not mention a key paper on the topic published in the same *Journal of Neurosurgery* 5 years earlier. This contribution by Ford and McLaurin in 1963[6] indicated that the more dura that is initially separated from the bone, the greater the force available to produce further separation of the dura. However, despite this finding, the authors considered most of an EDH was accumulated shortly after trauma. They argued that subsequent deterioration was not necessarily due to ongoing bleeding but to changes to the intracranial contents secondary to the initially formed hematoma.

THE HISTORY OF EVOLVING NOTIONS ABOUT EPIDURAL HEMATOMA FORMATION

There is no question that Bell's findings were crucial. Everyone since has agreed that for an EDH to form, the initiating trauma must cause dura to separate from the skull forming an epidural pocket. The subsequent increasing volume from ongoing bleeding from a meningeal artery remained a matter for debate. The next step was again the result of Erichsen's work. In 1895,[7] in the 10th edition of his surgery text book, he restated that the dura was separated from the skull at the time of initial trauma in agreement with Bell. However, he went further to describe a mechanism whereby bleeding may separate dura further from the skull. The greater the area of separated dura, the greater the amplification of the force acting across the dura for a given pressure (see Fig. 12.1).[7] Thus, whether secondary dura separation due an EDH volume increase from ongoing bleeding occurred or not, Erichsen had provided a mechanism whereby it would be possible. The question was however still not settled as the strengths of the necessary forces had not been adequately determined.

The next experimental work concerning dura separation was per- formed by Dr. Milroy Paul who in the only experiments to examine how much dura separation occurs following a blow to the head in humans, found that for a frontal blow, the area of dura separated was limited.[8] However, he noted that there was an extensive area within the skull where finger dissection could loosen the dura from the skull with relative ease. This was in fact not his own discovery. In a French anatomy text from 1900, there is a passage which can be translated as follows. "The osteo-dural adhesions are the weakest in the temporo- parietal region and in the occipital region. There is a special zone in which the dura mater is easily detached, not only by the anatomist's forceps, but also by the blood effusions which occur at this level as a result of a wound in the middle meningeal artery. This area, to which Marchant (Th. Paris, 1880) has given the name of 'detachable zone,' extends, from front to back, from the posterior border of the lesser wing of the sphenoid up to 2 or 3 centimeters from the internal occipi- tal protuberance; from the vicinity of the upper longitudinal sinus to the transverse line which joins the top of the lesser wings of the sphe- noid at the base of the rock. It measures, on average, 13 centimeters in length by 12 centimeters in height (Marchant)."[9] Thus, the combina- tion of the French book and Paul's work suggests that there is a region where dura separation is relatively easy and this area corresponds quite closely to the location of clinical EDHs.

However, the question remains. How much force is required. The answer to this question comes from two sources. The best known is the papers of Ford and Mclaurin have shown that an initial separation of half a cm^2 (the area of dura with a diameter of 8 mm) was adequate for further separation to occur.[6] Thus, it seems reasonable to believe that a ruptured meningeal artery can produce enough force to induce further separation of dura after trauma. While Ford and Mclaurin demonstrated that a meningeal artery could generate enough force to further detach the dura, they were skeptical to that happening. They believed that the prolonged clinical course followed by improvement or deterioration reflected what happened to the brain within the dura. This opinion was based on an anatomical study of the intracranial contents at different times after an experimentally induced hematoma. However, the variety of parameters in a small number of animals makes it difficult to be certain of the significance of the findings. Certainly, the findings did not provide evidence to exclude other mechanisms. Thus, it must be considered that if Paul was correct and the area of dura separated by a blow is limited to 3 or 4 square inches, then such further separation will be necessary to produce a clinical life-threatening EDH.

The other relevant source is Murzin and his colleagues in Moscow,[10] as indicated in Chapter 12, Developing Notions of Pathophysiology. They measured the extreme variability of the strength of attachment of the dura to the bone. This would contribute to the variability in the formation of EDHs seen in the clinic.

At this point, it is worth noting that an EDH can develop after a delay. In 1978, a paper was published in which a patient was shown on angiography to have no EDH on admission, 5 h after trauma. Following rehydration and clinical deterioration, a second angiogram demonstrated an EDH 8 h after trauma. This is obvious volume increase occurring after the initial trauma. However, it does not show whether an epidural pocket had been formed at the time of injury, though presumably it must have. The patient was hypotensive on admission and the EDH developed after rehydration made her normotensive. The pocket could have been there all along and only filled up when the blood pressure rose.

DURA SEPARATION EXAMINED IN OSLO

While Ford and Mclaurin had performed anatomical studies of the intracranial changes following an experimental EDH, there had never been a study of physiological parameters related to different degrees of dura separation. In view of this, a series of experiments were undertaken to examine just how different degrees of dura separation affect the response to an EDH.[11] The animal model was used is illustrated in

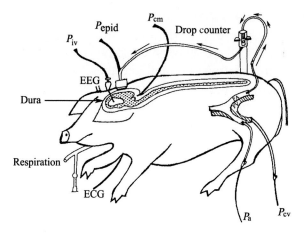

Figure 14.1 This shows the set up with the pig which was anesthetized, intubated with ECG, EEG, and respiration rate monitoring. A drop counter is placed between a femoral artery catheter and a metal bolt (18 or 30 mm diameter) screwed into the skull. The direction of blood producing the hematoma is indicated by the small arrows. FA, femoral artery; FV, femoral vein.

Attached to the inside of the bolt was a rubber balloon to collect the epidural blood. This was necessary because without the balloon the volume of bleeding would have been impossible to measure. The balloon was made of very thin rubber and it was checked no tension developed in its walls for the volumes of hematoma used in these experiments. Thus, balloon wall tension could not have interfered with the results.

This bolt permitted the measurement of epidural pressure and the introduction of blood simulating a hematoma. The other catheters are indicated which measured P_a = blood pressure, P_{epid} = pressure in the epidural space, P_{iv} = intraventricular pressure. P_{cm} = cisterna magna pressure, P_{cv} = central venous pressure.

The author is grateful to Springer Verlag for permission to use this image.[11]

Fig. 14.1. The model permitted the measurement of the following parameters.

Arterial pressure	P_a
Epidural pressure	P_{epid}
Intraventricular pressure	P_{iv}
Cisterna magna pressure	P_{cm}
Central venous pressure	P_{cv}
Bleeding volume and bleeding rate	
ECG and EEG	
Respiration	

It functioned well except for P_{cv} recording. This was never satisfactory as the catheter had a tendency to occlude from thrombosis, so that reliable recordings could not be made. One detail must be emphasized. This model did NOT permit assessment of progressive dura detachment but only the effect of different degrees of initial detachment.

Effect of Initial Dura Separation

The patterns of response to EDH are shown in Fig. 14.2. With slight dura separation (in these experiments 2.5 cm^2), the bleed is self-limiting, the final hematoma is small and there is no progressive rise in intracranial pressure (ICP). The difference between intraventricular pressure and cisterna magna pressure is slight and stable. It can be considered normal. There is no excessive rise in pressure at the

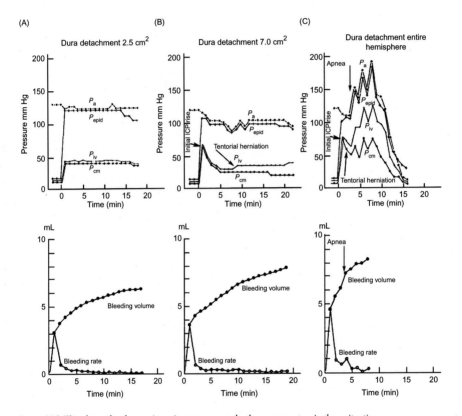

Figure 14.2 This shows the changes in various pressure and volume parameters in three situations.

(A) This shows a limited bleed of 6 cm^3 which is compensated for from the start. No further danger from delayed rises of ICP. The initial bleeding rate is high and falls rapidly according to the principles outlined in Chapter 13, Intracranial Vascular Dynamics.

(B) This shows a bleed of 7 cm^3 with an initial rise in ICPs which then fall again to a stable though raised level. Again, the bleeding rate is initially high and then falls. This is followed after a delay by a secondary rise in intraventricular pressure due to tentorial herniation. The P_{iv} continues to rise after bleeding has stopped.

(C) In this dura is detached over the entire hemisphere and within 8 min the hematoma volume is 8 cm^3. There is almost instant tentorial herniation. The supratentorial perfusion pressure falls to 50 mm Hg within a minute which is too low for adequate CBF. The animal stops breathing after 2–3 min and succumbs. A Cushing response ensues which maintains the bleeding until 8 min have passed. There is no stabilization of the ICP at a new higher level. The damage to the CBF and intracranial contents is too fast. This extreme situation is illustrative of what could happen but is unlikely to occur clinically (see text).

The author is grateful to Springer Verlag for permission to use this image.[11]

beginning of bleeding followed by a fall. The rise observed is easily compensated by the normal mechanisms in place to maintain ICP and cerebral blood flow (CBF). There are no secondary later rises. The pattern is one of an easily compensated small volume hematoma.

If the dura separation is greater (7 cm^2), there are three major differences. There is an initial rapid rise in intraventricular and cisterna magna pressures due to an accumulation of hematoma which is too rapid for normal compensation mechanisms, in keeping with the intracranial reaction to acute volume loading as outlined in Chapter 13, Intracranial Vascular Dynamics. However, these pressures rapidly fall again to a plateau which is still raised. This fall probably reflects the expulsion of blood and cerebrospinal fluid (CSF) from the intracranial compartment according to the principles of the Monro-Kellie doctrine. This plateau is succeeded by a secondary rise in intraventricular pressure with no concomitant rise in cisterna magna pressures. This must be due to obstruction of the continuity between the supratentorial compartment and the posterior fossa. It was considered that this was most likely due to tentorial herniation and this was subsequently shown to be the case as will be recounted in Chapter 15, Intracranial Effects of Epidural Bleeding. The intraventricular pressure continues to rise even after the cessation of bleeding ending in apnea and a Cushing response and the death of the animal, though that is not visible on the chart in Fig. 14.2. However, in a similar experiment, the pressures are shown leading to a lethal outcome as shown in Fig. 14.3.

With extensive dura separation over the entire hemisphere, a lethal hematoma accumulated rapidly followed by a Cushing response and death. There is no steady state as with the smallest hematoma and no stabilization after an initial rapid rise, as the rate of hematoma accumulation is too rapid for any compensation mechanism to be possible. Apnea occurs after 3 or 4 min and is followed by the hypertension and bradycardia typical of the Cushing response. This will be considered in more detail in Chapter 15, Intracranial Effects of Epidural Bleeding.

Thus, with varying degrees of dura separation, it was possible to demonstrate varying reactions from complete compensation for the presence of a small hematoma, via gradually overwhelmed compensation with a greater area of dura separation up to total failure of compensation with a wide separation.

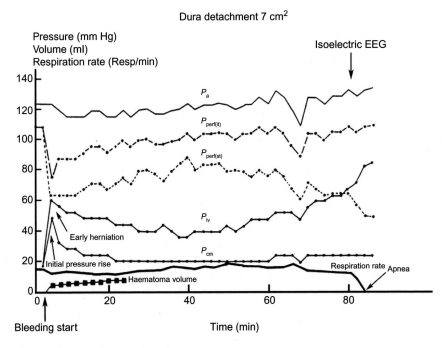

Figure 14.3 In this experiment with 7 cm^2 dura separation, the animal survived for roughly 80 min. There is the same rapid initial rise in ICP as before followed by stabilization at a raised level. The tentorial herniation in this case occurred earlier. The supratentorial and infratentorial pressures are derived and charted. The isoelectric EEG, signifying cessation of brain activity occurred when the supratentorial cerebral perfusion pressure fell below 60 mm Hg, as described in Chapter 13, Intracranial Vascular Dynamics. This occurs a short time before apnea occurs. A clear progressive rise in supratentorial pressure may be seen. The reasons for this are described in the next chapter.
The author is grateful to Springer Verlag for permission to use this image.[11]

The current work shows the pathophysiological reactions to separations of different degrees from speedy control of a limited bleed to rapid death. In clinical terms, the extreme reaction seen where the dura is separated over an entire hemisphere is an unlikely course. As demonstrated again by Paul, using blunt finger dissection, the area where the dura may be easily separated has limits equivalent to those observed at surgery. Total separation in practice does not occur. Thus, the likely clinical response is that seen in this series with a separation of 7.0 cm^2. It is important to note for later discussion in this chapter that in all the experiments the epidural pressure was at all times slightly less than but close to the systemic arterial pressure.

If the initial dura separation will tend to be limited, as indicated by Paul, then the resulting hematoma will need to expand to become lethal. The experimental design described separates the dura prior to the bleed. What the work reported here has shown is there are two other phenomena at work. There is most often a sharp rise in ICP initially which then falls to a lower though elevated level. If the bleed is of sufficient size, this is followed by a secondary rise in pressure which persists to death, most probably the result of tentorial herniation. In clinical terms, the early rise in ICP could contribute to an early loss of consciousness and the secondary rise to secondary loss of consciousness after a latent interval.

VEINS AND EPIDURAL HEMATOMA

Nobody knows to what extent venous blood can fill the epidural space created by a blow to the head which separates the dura from the skull. There is no way to examine this clinically as such an event would have occurred before the patient comes to the hospital. To date there is no experimental model which could see if it were possible. On the other hand, it is not possible for venous bleeding to cause the posttraumatic expansion of an EDH, as described in the case of arterial bleeding. The pressure of a bleeding source must be greater than the surrounding pressures. The pressure in the dural veins is downstream from that in the rest of the intracranial structures and thus the pressure of the blood in them is the lowest of any intracranial vessel. Moreover, Ford and McLaurin provided experimental evidence that bleeding from dural veins will never generate enough pressure to cause a progressive increase in volume.[6]

The earlier paragraph relates to situations where the initial ICP is normal. It is known, as recounted in Chapter 12, Developing Notions of Pathophysiology, that EDHs can form when CSF is drained too rapidly. In principal, venous bleeding could contribute to the formation of such hematomas. There is certainly no major artery involved. As also mentioned in Chapter 12, Developing Notions of Pathophysiology, such hematomas are uncommon and probably only occur in patients where the adhesion of the dura to the skull is unusually weak.

There are references in the literature to EDHs due to venous bleeding,[12-18] in some cases with active bleeding at operation. This gives the impression that ongoing bleeding is a cause of increasing hematoma volume. Yet from the arguments noted earlier, this should not be possible. In fact, there is a simple explanation. It is suggested that while venous bleeding cannot expand an EDH in the closed skull, at operation the skull is open. Any veins torn during trauma may well bleed once the closed box of the skull is opened. This does not mean that their bleeding could have expanded a hematoma prior to surgery.

With respect to EDHs associated with tears in dural sinuses and no other source of hemorrhage, it is suggested, albeit speculatively, that the hematoma filled an epidural pocket as mentioned earlier but that there could have been no subsequent increase in EDH volume once that pocket was filled.

ARTERIOVENOUS SHUNT

In the early 1980s, an attempt was made in Oslo to elucidate some of the unanswered questions with respect to the formation of EDHs. Among the questions asked were:

1. What are the physiological responses to EDH?
2. What are the differences between epidural and other intracranial bleeds?

Some of the answers to Question 1 have been outlined in the earlier section on dura separation. The approach to Question 2 was undertaken on dogs.[19] In the first of a series of studies, there were altogether nine different bleeding models. Only two are relevant in the present context. They both involved bleeding from the femoral artery in a way that was similar to that described in the pig in Fig. 14.1. However, the blood was taken from the artery via a drop counter (to assess the volume of bleeding) and into either the subarachnoid space via the cisterna magna or the epidural space. Both models were lethal because the animals and the tubing were heparinized so that bleeding continued until death. It was shown that subarachnoid bleeding was self-limiting in keeping with the pattern of bleeding shown in Fig. 13.6. It was associated with a marked rise in ICP.

On the other hand, the bleeding into the epidural space continued unabated for a long time and while the ICP rose, it did not initially rise as high as during the subarachnoid bleeding. The bleeding could persist for over 200 min as shown in Fig. 14.4.[19] The volume bled into the head far exceeded the intracranial volume before a final respiratory arrest and isoelectric EEG occurred. Thus, not all the blood can have stayed in the head but must have been diverted. The most likely explanation was that some of the blood was leaving the epidural space via veins. This notion gains weight from the observation that the epidural pressure is much lower than in the earlier experiments illustrated in Figs. 14.2 and 14.3.

In another set of experiments, dura was separated from the skull of dogs and saline containing a radio-opaque medium was injected into

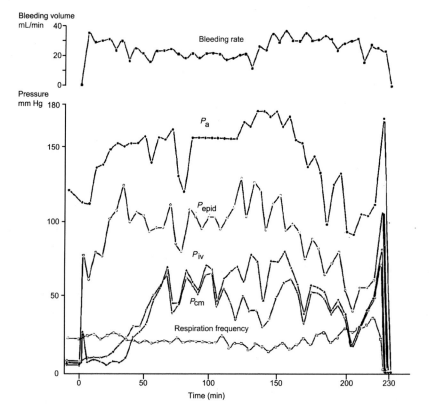

Figure 14.4 In this experiment, there are three major difference in the physiological response to bleeding compared to what was seen in the experiments illustrated in Fig. 14.3. The bleeding is taking place directly into the epidural space. First, the bleeding rate and volume, shown in the upper trace, far exceeds the intracranial volume. Second, the duration of the experiment was much longer than seen in Fig. 14.3 where bleeding took place into a closed balloon. Third, the epidural pressure is much lower than in the earlier experiments consistent with the escape of some of the blood from the epidural space. All of these factors indicate the probability of an arteriovenous shunt out of the epidural space.
The author is grateful to Springer Verlag for permission to use this image.[19]

the epidural space during ongoing experimental epidural bleeding. The medium was returned to the neck veins via diploic veins.[20] An arteriovenous shunt was thereby demonstrated. This means that during epidural bleeding while some blood may remain in the epidural space much of it can be shunted out again. This phenomenon permits a maintained low ICP and high bleeding volume. In a clinical context, such a shunt could be the mechanism underlying the lucid interval between the injury causing an EDH and the secondary deterioration.

Finally, the preoperative angiograms of 35 patients were examined for signs of a shunt.[21] Twenty-one patients had extravasation of contrast medium from the middle meningeal arteries. Of these, 20 had a bleeding middle meningeal artery at surgery. Seventeen of these 21 patients had shunting of contrast to meningeal or diploic veins. Thus, the presence of an arteriovenous shunt had been demonstrated both in canine experiments and human patients. Regrettably, the existence of this arteriovenous shunt has been totally ignored in the literature until 2015. A case report from Japan was published in "Neurosurgery." It is stated in the abstract that "Intraosseous dural arteriovenous fistulae (DAVF) are rare, especially those with drainage into the diploic venous system. The clinical presentation depends on the location of the lesion. This is the first report of an intraosseous DAVF associated with acute epidural hematoma."[22] It is most improbable that there was deliberate error in this case. It is almost certain that this statement could be made because the journal in which the Oslo work was published is expensive and not universally followed outside Europe.

CONCLUSION

The work described in this chapter concerns the formation of EDHs. It has previously been known that initial dura separation is a prerequisite for the formation of a hematoma. It would seem that the necessary dura separation was probably limited in extent and that there is a wide area within the skull where the variable dura adhesion to the skull was relatively low. This area was the location of clinical EDHs. The work described here demonstrated how different degrees of separation affect the course of the bleed and the brain's reaction to it.

The role of veins in EDH has been discussed. Venous bleeding may result in filling a pocket of dura separation at the time of injury but it seems impossible for it to produce further separation of dura from the skull. Finding venous bleeding at surgery says little about the process of bleeding since what happens in the closed and open skull would be different. Veins on the other hand have an important function in draining blood out of the epidural space via a shunt. The pressure changes shown in these experiments demonstrate an initial quick rise in ICP which rapidly falls to a lower, though still raised level. The arteriovenous shunt could undoubtedly lead to the persistence of this plateau at a lower level. However, Ford and Mclaurin suggested that after an initial bleed, subsequent changes were the result of changes in the brain itself. The experiments in the current work indicating the gradual development of a pressure gradient across the tentorium support this notion.

Thus, it would seem that the degree of dura separation has a key influence on the size of an EDH. However, it would also seem that the arteriovenous shunt and changes within the brain could both contribute to the clinical course with its classical unconsciousness, lucid interval, and secondary loss of consciousness.

REFERENCES

1. Bell C. *Surgical Observations*. London: Longmans; 1816.

2. Abernethy J. *Surgical Observations on Injuries of the Head*. London: Longman; 1810.

3. Gallagher JP, Browder EJ. Epidural hematoma. Experience with 167 patients. *J Neurosurg*. 1968;29:1–12.

4. Erichsen J. Lectures on Injuries of the Head. *Lancet*. 1878;5:1–4.

5. Zander E, Campiche R. Extradural hematoma. In: Krayenbühl H, ed. *Advances and Technical Standards in Neurosurgery. 1: Springer-Verlag, Wien*. 1965:121–139.

6. Ford LE, McLaurin RL. Mechanisms of extradural hematomas. *J Neurosurg*. 1963;20:760–769.

7. Erichsen J. *The Science and Art of Surgery*. 10th ed London: Longman and Green; 1895.

8. Paul M. Haemorrhages from head injuries. *Ann R Coll Surg Engl*. 1955;17(2):69–101.

9. Testut L. *Traité D'Anatomie Humaine*. Paris: Octave Doin; 1900.

10. Murzin VE, Goriunov VN. Study of the strength of the adherence of the dura mater to the bones of the skull. *Zh Vopr Neirokhir Im N N Burdenko*. 1979;4:43–47.

11. Ganz JC, Zwetnow NN. Analysis of the dynamics of experimental epidural bleeding in swine. *Acta Neurochir (Wien)*. 1988;95(1-2):72–81.

12. Stevenson GC, Brown HA, Hoyt WF. Chronic venous epidural hematoma at the vertex. *J Neurosurg*. 1964;21:887–891.

13. Prat R, Galeano I. Posterior fossa venous epidural hematoma. Based on 2 cases. *Neurologia*. 2003;18(1):38–41.

14. Singh S, Ramakrishnaiah RH, Hegde SV, Glasier CM. Compression of the posterior fossa venous sinuses by epidural hemorrhage simulating venous sinus thrombosis: CT and MR findings. *Pediatr Radiol.* 2016;46(1):67−72.

15. Kissel P, Boggan JE, Wagner Jr. FC. CT evolution of an acute venous epidural hematoma. *J Emerg Med.* 1989;7(4):365−368.

16. Khwaja HA, Hormbrey PJ. Posterior cranial fossa venous extradural haematoma: an uncommon form of intracranial injury. *Emerg Med J.* 2001;18(6):496−497.

17. Samandouras G. *The Neurosurgeron's Handbook*. Oxford: Oxford University Press; 2010:208.

18. Rowbotham GF. *Acute Injuries of the Head*. Edinburgh & London: E & S LIvingstone; 1964.

19. Zwetnow NN, Habash AH, Lofgren J, Hakanson S. Comparative analysis of experimental epidural and subarachnoid bleedings in dogs. *Acta Neurochir (Wien).* 1983;67(1-2):67−101.

20. Habash AH, Zwetnow NN, Ericson K, Lofgren J. Arterio-venous epidural shunting in epidural bleeding radiological and physiological characteristics. An experimental study in dogs. *Acta Neurochir (Wien).* 1983;67(3-4):291−313.

21. Habash AH, Sortland O, Zwetnow NN. Epidural haematoma: pathophysiological significance of extravasation and arteriovenous shunting. An analysis of 35 patients. *Acta Neurochir (Wien).* 1982;60(1−2):7−27.

22. Yoshioka S, Kuwayama K, Satomi J, Nagahiro S. Transarterial N-Butyl-2-cyanoacrylate embolization of an intraosseous dural arteriovenous fistula associated with acute epidural hematoma: technical case report. *Neurosurgery.* 2015;11(Suppl. 3):E468−E471.

CHAPTER 15

Intracranial Effects of Epidural Bleeding

PHYSIOLOGICAL CHANGES
Outcome

Chapter 14, Factors Affecting the Formation of Epidural Hematomas, indicated the effects of physiological arteriovenous shunting and dura detachment on the development of an epidural hematoma (EDH). It is now necessary to examine how such a hematoma affects the intracranial contents and how some of them become lethal while others do not. It was not of course possible to examine the clinical neurological condition of the animals employed. Analysis was based on the effects of bleeding on respiration, arterial pressure, intracranial pressures (ICPs), hematoma volume, cerebral blood flow (CBF), and anatomical changes. The first of these to be considered is respiration.

RESPIRATION

The question arises about what distinguishes lethal from survivable bleeds. All pigs were prepared as outlined in Chapter 14, Factors Affecting the Formation of Epidural Hematomas, with a dura detachment over an area with a diameter of 30 mm. This model was chosen as some animals survived and some succumbed. All animals with a smaller dura detachment survived while all animals with a wider dura detachment died in less than 5 min. The 30-mm detachment permitted examination of the differences between experiments where the animal succumbed and those in which they survived. The sign of irreversible death was an isoelectric electroencephalogram (EEG). It became clear that persisting respiration was important in this process and that isoelectric EEG was preceded by apnea. Comparisons were made between animals which breathed spontaneously with those where respiration was maintained by artificial ventilation. Bleeding volumes were similar in all the experiments so that this parameter did not affect the results.[1]

All the spontaneously breathing animals succumbed. A typical chart is shown in Fig. 15.1. As described in Chapter 14, Factors Affecting

Intracranial Epidural Bleeding. DOI: https://doi.org/10.1016/B978-0-12-812159-7.00015-1

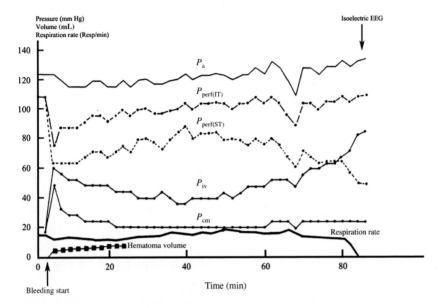

Figure 15.1 This spontaneously breathing animal died following epidural bleeding. It may be seen that the respira-tion rate began a gradual fall a few minutes after the first hour of the experiment, a fall which accelerated until apnea after just over 80 min. There is a terminal hypertension, together with the apnea; in other words, a Cushing's response.

P_a = *arterial pressure,* $P_{perf\ (IT)}$ = *infratentorial perfusion pressure,* $P_{perf\ (ST)}$ = *supratentorial perfusion pressure,* P_{iv} = *intraventricular pressure, and* P_{cm} = *cisterna magna pressure.*

The author is grateful to Springer Verlag for permission to use this image[1].

the Formation of Epidural Hematomas, after cessation of bleeding there was rise in supratentorial pressure without a concomitant rise in posterior fossa pressure. In the presence of rising supratentorial pressure, the animal's respiration rate slows and stops and soon after the EEG became isoelectric. This pattern was consistent. The only difference between different experiments with spontaneous respiration was the speed of the process. In every animal, there was a gradual secondary rise in supratentorial pressure starting after the cessation of bleeding. This was the same as described in Chapter 14, Factors Affecting the Formation of Epidural Hematomas. It was a very consistent process. There remained the question of whether the falling respiration rate was due to the pressure changes or contributed to them. The timing of the changes shown in Fig. 15.1 would suggest that respiration changes followed changes in ICP rather than the reverse but it was not completely clear.

Fig. 15.2 illustrates the course with artificially assisted ventilation and a lethal outcome. Once again there was the gradual secondary rise in supratentorial pressure starting after the cessation of bleeding. Yet again there was no concomitant increase in posterior fossa pressure. This rise in P_{iv} led again to a gradual fall in $P_{perf(ST)}$ to levels inconsistent with an adequate CBF. This sequence had nothing to do with respiration as the animal was being artificially ventilated. Thus, it would seem that in the relationship between ICP and respiration, it was the rise in ICP which led to hypoventilation and not the other way round. In the chart for a lethal ventilated experiment as soon as the animal was taken off the ventilator, respiration did not restart and the EEG rapidly became isolectric with death following.

The course with animals which survived was quite different in one single respect. There was no secondary rise in supratentorial pressure following the cessation of bleeding. Following the disconnection of the

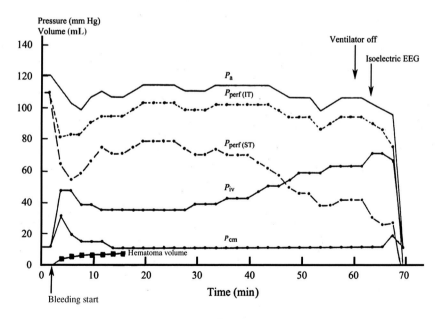

Figure 15.2 This chart shows exactly the same secondary rise in P_{iv} with a rising gradient between P_{iv} and P_{cm} and falling supratentorial perfusion pressure as in Fig. 15.1. Respiration was not restarted after disconnection of the ventilator. It would seem hypoventilation was secondary to the pressure changes and not their cause. The author is grateful to Springer Verlag for permission to use this image[1].

animal from the ventilator, there was a short rise in arterial pressure and a concomitant rise in ICPs while respiration was reestablished.

The findings suggest that survival required the absence of the secondary rise in ICP. Increasingly, it looked as though the crucial process leading to death was tentorial herniation but more information was required. Some of this was physiological and some of it was anatomical. The crucial physiological changes were the distribution of regional cerebral blood flow (rCBF) during bleeding.

CEREBRAL BLOOD FLOW AND EPIDURAL BLEEDING

rCBF was examined using a microsphere technique in seven lethal epidural bleeds with artificial ventilation. The dura was detached over the entire hemisphere to ensure a consistently lethal outcome.[2] The charts in all prior experiments had shown a sharp initial rise in ICPs followed by a steady state which eventually would change in lethal cases to a secondary rise. Thus, CBF measurements were taken prior to bleeding, at the time of the steady state and after the EEG became isoelectric (see Fig. 15.3). In every case stage, the autoregulatory and CO_2 responsiveness of the cranial circulation was tested prior to the bleed. It was clear that there was no significant change in rCBF at the time when the ICPs had reached a plateau following a fall after the initial sharp rise. Checking cerebrovascular resistance demonstrated significant changes in all regions indicating that autoregulation was functioning. When the EEG had become isoelectric, there was a reduction in rCBF in all regions but it was much more marked in the cortex above the tentorium directly under the EDH and in the upper midbrain; a location consistent with compression due to tentorial herniation. The lower midbrain, and pons also showed ischemia but the medulla was relatively spared and the rCBF values were consistent with cell survival and continuing function. The pressure course was similar to that shown in Fig. 15.4 where there was no Cushing response which on the other hand was seen in Fig. 15.1. This will be discussed further in Chapter 16, Terminal Changes in Epidural Bleeding.

CAUSE OF DEATH

Earlier work had been done to determine the mechanisms underlying the Cushing Response.[3] Certain relationships were determined using a

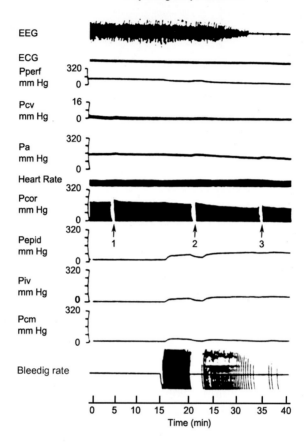

Figure 15.3 Physiological changes observed in an animal which suffered a lethal EDH, continuous recording. CBF measurement was done at three stages.
The author is grateful to Springer Verlag for permission to use this image[2]
1. Prior to bleeding.
2. When initial ICP rise has leveled out to a plateau.
3. Just after the EEG became isoelectric.

rubber balloon expanded at a constant rate. Roughly speaking if the balloon achieved a volume of 10% of the intracranial volume, death ensued. This was called the apnea volume. When the balloon reached approximately 5% of the intracranial volume, changes in physiological parameters occurred with bradypnea, bradycardia, and increasing blood pressure. These changes occurred at the same time as tentorial herniation. It should be noted that the volume loading was not self-limiting being produced by an external mechanical source.[3,4]

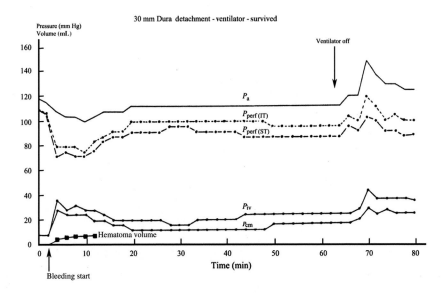

Figure 15.4 This animal survived the hemorrhage and survived disconnection of the ventilator. Cerebral perfusion pressure was maintained. No secondary rise in P_{iv} observed.
The author is grateful to Springer Verlag for permission to use this image[1].

In the current context, the volume loading was via epidural bleeding into a rubber balloon. This was self-limiting being produced by the animal itself. In these experiments, a tentorial herniation appeared to occur in all lethal cases. The volume of the lethal hematomas was just under 10% of the intracranial volume, so that the volume tolerance of an EDH was very similar to that of an intracranial balloon.

What is of more importance is the distribution of ischemia. In the balloon expansion experiments, there was a gradient of ischemia from above downwards to the lower pons with sparing of the medulla. The distribution of ischemia in epidural bleeding was very similar. The consequence of this finding is that foramen magnum herniation is not a requirement for a lethal bleed terminating with a Cushing Response. It remains to see if the anatomical changes observed during epidural bleeding match the above interpretation.

ANATOMICAL CHANGES

The only available way to study anatomical changes during epidural bleeding is by using MRI. Changes were compared between lethal and nonlethal bleeds. The pattern of pressure changes was the same as in earlier experiments. There was a rapid initial rise in early ICPs which fell to a plateau. An initially stable pressure gradient developed between the ventricle and the cisterna magna. The perfusion pressure remained adequate. Six animals survived, two did not.

In the survival experiments, there was no change in hematoma volume after the first few minutes. See Fig. 15.5. The ipsilateral lateral ventricle, third ventricle and to a much lesser extent contralateral lateral ventricle suffered mild compression and displacement. There was compression but not occlusion of the cisterna ambiens around the midbrain and the cerebral aqueduct. There was also minor compression of the cisterna magna. There was an increase in the white matter T2 signal indicating accumulation of water which is most pronounced ipsilateral to the bleeding. There were no late anatomical changes thus an anatomical steady state developed which paralleled the above-mentioned physiological steady state.

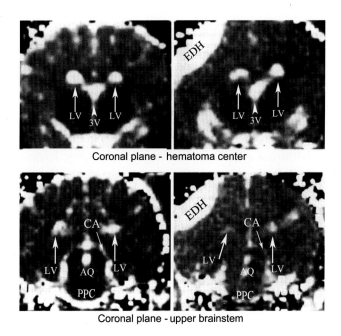

Coronal plane - hematoma center

Coronal plane - upper brainstem

Figure 15.5 MRI T2 sequences in an animal which survived.
LV, *lateral ventricle;* 3V, *third ventricle;* AQ, *aqueduct;* PPC, *pre pontine cistern;* EDH, *epidural hematoma;* CA, *cisterna ambiens, which is partially closed in the lower right image.*
On the left before bleeding, on the right after. There is minor shift of the LVs and third ventricle. The ipsilateral ventricle is a little compressed. The PPC is partly occluded.

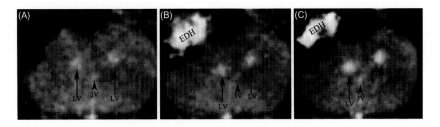

Figure 15.6 MRI T2 weighted FLASH sequences in an animal which succumbed.
(A) Before bleeding. (B) Just after bleeding stopped. (C) Just before terminal collapse.
LV, *lateral ventricle;* 3V, *third ventricle;* EDH, *epidural hematoma.*
(B) The entire ventricular system is displaced toward the opposite side. EDH volume is maximal.
(C) The lateral displacement of the ventricles is less. The ventricles have begun to dilate. The EDH is thinner perhaps being compressed by the secondary hydrocephalus.

The lethal experiments showed quite a different picture. See Fig. 15.6. The tolerance of the intracranial space to volume loads had been the subject of earlier work. The key paper suggested that an acute volume load of around 10% was consistently lethal.[4] In the two lethal experiments in this study the hematoma volume was rather less than this but there was an explanation consistent with the earlier findings. The supratentorial perfusion pressure was lower. The shifts of the ventricles were more marked. The cisterna ambiens and aqueduct were not compressed but occluded. The frequently noted secondary rise in ICP was noted but it was accompanied by some unexpected anatomical findings. The ventricles started to redilate suggesting the development of a secondary hydrocephalus due to the herniation induced occlusion of the cerebrospinal fluid pathways mentioned earlier. In view of the Monro-Kelly Doctrine, this could only happen if some other structure left the supratentorial compartment. The observation of a gradual progressive closure of the cisterna magna indicates this explanation is consistent with the observations. Another source of volume loading was the increase in tissue water indicated by increasing T2 signal. This remained mild in the posterior fossa and it would seem that medullary compression was not a component responsible for a lethal outcome.

CONCLUSIONS

Epidural bleeding may or may not be lethal. The outcome is largely related to the hematoma volume. However, in laboratory experiments, it is easier to apply a more precise control to the parameters being recorded than is possible in the clinic. Thus, while laboratory animals seem not to tolerate an acute volume loading in excess of 10% of intracranial volume, this author has observed EDHs in patients which far exceeded this volume. In one particular case, the patient was a young drug addict whom over the course of a number of years suffered two epidural hemorrhages, one on one side and one on the other. The volume was measured on computed tomography and was over 200 cm^3 on both occasions which is far above the 10% of the intracranial volume of most adult human males. A similar phenomenon has also been recorded in the literature.[5] The explanation may perhaps be the result of EDH in patients with cerebral atrophy. Whatever the volume range of the reaction volume and apnea volume in humans be, there is no reason to consider that the principles of cerebral reaction are essentially different from those observed in the laboratory.

The key components of a lethal EDH would seem to be first tentorial herniation. Thereafter, aqueduct occlusion can induce a secondary hydrocephalus and the associated cerebral ischemia is associated with a cerebral edema. Thus, the volume loading in these lethal cases consists of the volume of the hematoma and the volume of extra water accumulated in the ventricles and the cerebral parenchyma. The sum of these volume loads causes a degree of cerebral ischemia which is incompatible with life. However, it would seem that the ischemia does not extend as low as the medulla oblongata and that foramen magnum herniation is not a requisite for a lethal outcome. The sequence of physiological reactions occurring prior to death are considered in Chapter 16, Terminal Changes in Epidural Bleeding.

REFERENCES

1. Ganz JC, Zwetnow NN. A quantitative study of some factors affecting the outcome of experimental epidural bleeding in swine. *Acta Neurochir (Wien)*. 1990;102(3-4):164−172.

2. Ganz JC, Hall C, Zwetnow NN. Cerebral blood flow during experimental epidural bleeding in swine. *Acta Neurochir (Wien)*. 1990;103(3-4):148−157.

3. Schrader H. *Dynamics of Intracranial Expanding Masses. An Experimental Study with Particular Reference to the Cushing Response*. Oslo: Oslo University; 1985.

4. Zwetnow N, Schrader H, Löfgren J. Effects of continuously expanding intracranial lesions on vital physiological parameters. An Experimental Animal Study. *Acta Neurochir (Wien)*. 1986;80(1-2):47−56.

5. Paterniti S, Falcone MF, Fiore P, Levita A, La Camera A. Is the size of an epidural haematoma related to outcome? *Acta Neurochir (Wien)*. 1998;140(9):953−955.

CHAPTER 16

Terminal Changes in Epidural Bleeding

INTRODUCTION

According to Cushing's original description, the response had the purpose of reestablishing cerebral circulation when a rise in intracranial pressure (ICP) had reached the level of the arterial pressure. It consisted of three components: hypertension, bradycardia, and bradypnea. He provided evidence that the stimulation for hypertension originated in the vasomotor center, passed through the spinal cord to the sympathetic nervous system, and increased extracranial peripheral resistance. The bradycardia was affected via the vagus nerves. Bradypnea was the result of ischemia of the respiratory center in the medulla. Cushing's notion that the hypertension served the purpose of reestablishing circulation was soundly based as he observed improvements in circulation following each increase in blood pressure. These observations were made through a window in the cranium where pallor of the brain was seen when the ICP matched the arterial pressure. This improved following the subsequent increase in blood pressure.

He increased the ICP in two ways. In one set of experiments, the fluid was introduced into a balloon placed in the epidural space. To avoid the distortions produced by such a technique, he also infused the subarachnoid space producing a diffuse rise in ICP. Cushing's description of the CR is expressed with clarity and elegance both of language and experimental method. The importance of understanding the pathophysiology of the CR is that such understanding could improve the management of patients with acute raised ICP and close to death. However, Cushing's own work and the work of most subsequent researchers includes nonphysiological error. The increases in ICP are produced by an external source not subject to the regulatory mechanisms of the body.[1−15] Analysis of the components of the CR relates to experiments of this kind and is considered next.

Intracranial Epidural Bleeding. DOI: https://doi.org/10.1016/B978-0-12-812159-7.00016-3

THE CUSHING RESPONSE WHERE THE SOURCE OF INTRACRANIAL PRESSURE RISE WAS EXTERNAL

During a series of experiments undertaken to examine different aspects of the pathophysiology of intracranial epidural bleeding, it became apparent that certain aspects of our understanding of the physiology of the Cushing response (CR) needed to be reconsidered. It has been repeatedly described, both in association with a diffuse increase in intracranial pressure, using fluid infusion[3,5,14–19] and with a focal increase in pressure using a balloon.[2,4,10,12–14,20,21] A very similar response sequence has also been seen in association with hypoxia,[5,22] hypercarbia,[5,22] and ischemia.[5,22–24] The anatomical substrates of the three components of the CR are known to be separate. The respiratory disturbance was initially considered to be the result of ischemia of the so-called respiratory center in the medulla oblongata.[1] Evidence has more recently indicated that it follows ischemia of the so called "pneumotaxic center," in the pons.[13] Cushing and others considered that the bradycardia is mediated by the vagus,[1,25] a view confirmed by its absence following vagotomy. It can be elicited by stimulation of a discrete area in the neighborhood of the nucleus ambiguous in the medulla oblongata.[21,25] The vasopressor component is still considered to be elicited from the pressor area of the medulla[20,21,26] in a region lying between the inferior olive and the nucleus of the facial nerve.[20,21]

Over the years, there has been much disagreement as to the nature of the adequate stimulus for the elicitation of the CR. Some authors favored a direct effect on the brain stem either by mechanical distortion,[19] or by local pressure.[20,21] However, in a crucial set of studies, Schrader et al.[12–14] showed that pons ischemia was consistently associated with the CR while the effects of mechanical distortion and local pressure on the brain stem were inconsistent and variable. This was the first work which involved measurements of regional cerebral blood flow during an experimental Cushing's response. A supratentorial balloon expansion was used. The key finding was that *normal* medullary blood flow was maintained following a supratentorial expansion which caused a pressor response[27] and an isoelectric electroencephalogram (EEG).[10,27] Thus, some other mechanisms than medullary ischemia must trigger the pressor response and bradycardia in this situation. Schrader et al. also noted that hypotension increased the lethality of raised ICP, while hypertension reduced it thereby lending support to Cushing's notion that the hypertension had a beneficial effect on brain stem perfusion.

The Schrader work indicated that the respiratory component of the CR could involve the pneumotaxic center in the pons. The bradycardia was assumed to have its origin in the medulla. However, since it occurred before the medulla became ischemic, the mechanism underlying its elicitation remained unclear. The same is applied to the hypertensive response, the origin of which lies within the vasomotor center of the medulla oblongata. There was speculation in one of Schrader's papers that hypoxia might play a part in the CR but this was not pursued.[12]

THE CUSHING RESPONSE WHERE THE SOURCE OF INTRACRANIAL PRESSURE RISE WAS INTERNAL

During the course of the experiments, the results of which were described in this chapter and Chapter 15, Intracranial Effects of Epidural Bleeding, it was noted that in the presence of lethal bleeds, the use of the ventilator had a marked effect on the pressor response component of the CR.

The present remarks are only concerned with the elicitation of the response in association with a supratentorial expanding lesion, the origin of which was blood from a femoral artery diverted into a balloon in the epidural space in the head. Some of these animals breathed spontaneously, while some were artificially ventilated. The type of ventilation had no effect on hematoma volume, time to isoelectric EEG, or CPP at isoelectric EEG. There was nonetheless a marked difference in respect of the course of the systemic arterial pressure in the two situations, see Figs. 16.1 and 16.2. Thus, while the isoelectric EEG was associated with all the components of a CR in the spontaneously breathing animals, the hypertension was absent at isoelectric EEG, when mechanical ventilation was employed. The absence of a pressor response when mechanical ventilation is used is not an entirely new finding. Nagao et al. "noted such a response in only 16 of 30 ventilated cats."[10] Gonazales et al. found the response to be attenuated with the use of a ventilator in dogs.[4] As noted in the introduction, the response is known to be mediated by sympathetic nervous system neurons with their origin in the pressor area of the medulla oblongata. The consistent pressor response in spontaneously ventilating animals, in the present experience together with its absence in mechanically ventilated animals, indicates the importance of a respiratory disturbance for the

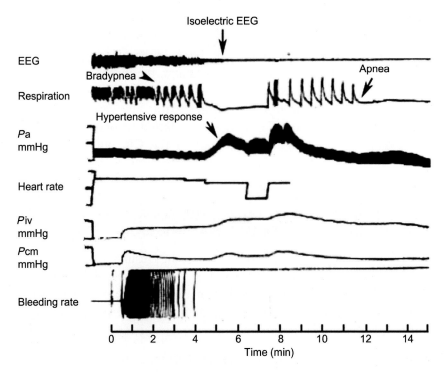

Figure 16.1 Lethal bleed with spontaneous respiration.
There is a rapid bleeding rate associated with a rapid rise in intracranial pressures. Note the typical Cushing response with bradypnea followed by hypertension and bradycardia, with subsequent cardiovascular collapse. The respiratory disturbance comes first. It is associated with the developing gradient between pressures above and below the tentorium. At no point does the ICP approach the blood pressure.

pressor response to occur. The importance of this disturbance is further underlined by the observation of a pressor response, following the disconnection of the ventilator, in the mechanically ventilated animals. In all animals at the time of the response, there was marked arterial hypoxia.

These findings are in keeping with Schrader's work where disturbances of respiration always preceded the pressor response, when a supratentorial expansion was employed.[14] This sequence has been confirmed in subsequent work, using epidural bleeding as the expanding lesion.[28] Moreover, it has been shown that with a supratentorial expanding lesion, an ischemic front extending from above the cerebral hemispheres to the pons is present at the time of apnea.[14] However, the medulla is spared from this ischemia. Sparing of the medulla was also seen in lethal experiments following epidural bleeding.[17] Thus, it

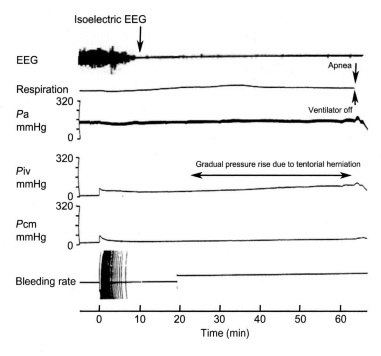

Figure 16.2 Lethal bleed with artificial ventilation.
Rapid bleeding rate and change in intracranial variables similar to Fig. 16.1 with isoelectric EEG after a few minutes. There was a slight bradycardia at isoelectric EEG but total absence of a hypertensive response. This response only occurred after the ventilator was disconnected and hypoxic blue arterial blood was visible. At no point does the ICP approach the blood pressure.

is logical that some other mechanism than medullary ischemia alone is responsible for the observed pressure response, in this situation. Gross hypoxia was present during the CR, in the spontaneously ventilated animals, as indicated by the color of the blood in the tubing of the bleeding system. This hypoxia is assumed to be a component of the asphyxia produced by the respiratory disturbance with which it was associated. It is known that hypoxia and hypercarbia can elicit a pressor response from the pressor area of the medulla.[5] Gross hypoxia was also present in the arteries of the artificially ventilated animals as soon as the ventilator was disconnected. This lends support to the notion of the importance of hypoxia or asphyxia in initiating a CR.

The results of this work have both theoretical and practical significance. In most experimental works, cited in the introduction, the ICP was raised to unphysiological levels never occurring in life; with ICP levels approaching the diastolic blood pressure. This did not occur in any of the epidural bleeds. The ICP was indeed markedly raised but always remained 40–50 mm Hg less than the blood pressure. Even so Cushing's responses were seen in all these experiments as soon as hypoxia was established. At the same time, the CBF to the medulla was consistently maintained at normal values. Moreover, foramen magnum herniation did not occur so that direct mechanical compression of the medulla was not a feature of the experiments. The sequence was always hypoxia due to respiratory disturbance followed by hypertension and bradycardia. Thus, it would seem that hypoxia was the key event to start a CR.

The practical consequences of the work are simple and direct. The significance of maintained respiration underlines the need for intubation and ventilation of patients rendered unconscious by a traumatic brain injury. This is particularly important because hypoventilation can lead to asphyxia which would dilate vertebral vessels and increase the intracranial volume load. Moreover, the observation in other work that a low blood pressure reduces the capacity of the brain to tolerate an intracranial volume load underlines the need to maintain intravenous fluids in head injured patients and to prevent hypotension.[29]

REFERENCES

1. Cushing H. Some experimental and clinical observations concerning states of increased intracranial hypertension. *Am J Med Sci.* 1902;124:375–400.

2. Hekmatpanah J. The sequence of alterations in the vital signs during acute experimental increased intracranial pressure. *J Neurosurg.* 1970;32:16–20.

3. Brashear RE, Ross JC. Hemodynamic effects of elevated cerebrospinal fluid pressure: alterations with adrenergic blockade. *J Clin Invest.* 1970;49:1324–1333.

4. Gonzales N, Overman J, Maxwell JA. Circulatory effects of moderately and severely increased intracranial pressure in the dog. *J Neurosurg.* 1972;36:721–727.

5. McGillicuddy JE, Kindt GW, Raisis JE, Miller CA. The relation of cerebral ischemia, hypoxia and hypercarbia in the Cushing respons. *J Neurosurg.* 1978;48:730–740.

6. Shalit MN, Covert S. Interrelationship between blood pressure and cerebral blood flow in experimental intracranial hypertension. *J Neurosurg.* 1974;40:34–38.

7. Fitch W, McDowall DG, Keaney NP, Pickerodt WA. Systemic vascular responses to increased intracranial pressure. 2. "Cushing" respons in the presence of intracranial space-occuyping lesions: systemic and cerebral hemodynamic studies in the dog and baboon. *J Neurol Neurosurg Psychiatry.* 1977;40:843–852.

8. Richardson TW, Fermoso JD, Pugh GO. Effect of acutely elevated intracranial pressure on cardiac output and other circulatory factors. *J Surg Res.* 1965;5:318–322.

9. Ducker TB, Simmons RL, Anderson RW. Increased intracranial pressure and pulmonary edema. Part 3: The effect of increased intracranial pressure on the cardiovascular hemodynamics of chimpanzees. *J Neurosurg.* 1968;29:475–483.

10. Nagao S, Sunami N, Tsutsui T, et al. Acute intracranial hypertension and brain-stem blood flow: an experimental study. *J Neurosurg.* 1984;60:566–571.

11. Schrader H. *Dynamics of Intracranial Expanding Masses. An Experimental Study with Particular Reference to the Cushing Response.* Oslo: Oslo University; 1985.

12. Schrader H. Regional cerebral blood flow and CSF pressures during the Cushing response induced by an infratentorial expanding mass. *Acta Neurol Scand.* 1985;72:273–282.

13. Schrader H, Hall C, Zwetnow N. Effects of prolonged supratentorial mass expansion on regional blood flow and cardiovascular parameters during Cushing response. *Acta Neurol Scand.* 1985;72:283–294.

14. Schrader H, Zwetnow N, Löfgren J. Regional cerebral blood flow and CSF pressures during Cushing response induced by supratentorial expanding mass. *Acta Neurol Scand.* 1985;71:453–463.

15. Zwetnow N, Schrader H, Löfgren J. Effects of continuously expanding intracranial lesions on vital physiological parameters. An experimental animal study. *Acta Neurochir (Wien).* 1986;80(1-2):47–56.

16. Brown FK. Cardiovascular effects of raised intracranial pressure. *Am J Physiol.* 1956;185:510–514.

17. Ganz JC, Hall C, Zwetnow NN. Cerebral blood flow during experimental epidural bleeding in swine. *Acta Neurochir (Wien).* 1990;103(3-4):148–157.

18. Häggendal E, Löfgren J, Nilsson NJ, Zwetnow N. Effects of varied cerebrospinal fluid pressure on cerebral blood flow in dogs. *Acta Physiol Scand.* 1979;79:272–279.

19. Thompson RK, Malina S. Dynamic axial brain-stem distortion as a mechanism explaining the cariorespiratory changes in increased intracranial pressure. *J Neurosurg.* 1959;16:664–675.

20. Doba N, Reis D. Localization in the lower brain stem of a receptive area mediating the pressor response to increased intracranial pressure. *Brain Res.* 1972;47:487–491.

21. Hoff JT, Reis DJ. Localization of regions mediating the Cushing response in the CNS of cat. *Ann Neurol.* 1970;23:228–240.

22. Downing SE, Mitchell JE, Wallace AG. Cardiovascular responses to ischaemia, hypoxia and hypercapnia of the central nervous system. *Am J Physiol.* 1963;204:881–887.

23. Dampney RAL, Kumada M, Reis DJ. Central nervous mechanisms of the cerebral ischaemic response. Characterizaton. Effect of brainstem and cranial nerve transection. *Circ Res.* 1979;45:48–62.

24. Kumada M, Dampney RAL, Reis DJ. Profound hypotension and abolition of the vasomotor component of the cerebral ischemic resons produced by restricted lesions of the medull oblongata in rabbit. Relationship to the so-called vasomotor center. *Circ Res.* 1979;45:63–70.

25. Thomas MR, Calarescu F. Localization and function of medullary sites mediating vagal bradycardia. *Am Physiol.* 1974;226:1344–1349.

26. Forster FM. The role of the brain stem in arterial hypertension subsequent to intracranial hypertension. *Am J Physiol.* 1943;139:347–350.

27. Ganz JC, Zwetnow NN. Analysis of the dynamics of experimental epidural bleeding in swine. *Acta Neurochir (Wien).* 1988;95(1-2):72–81.

28. Ganz JC, Zwetnow NN. A quantitative study of some factors affecting the outcome of experimental epidural bleeding in swine. *Acta Neurochir (Wien).* 1990;102(3-4):164–172.

29. Schrader H, Löfgren J, Zwetnow N. Influence of blood pressure on tolerance to an intracranial expanding mass. *Acta Neurol Scand.* 1985;71:114–126.

CHAPTER 17

Status Quo Vadis

STATUS QUO

So, after 16 chapters, we have some information on which to base some conclusions.

1. It took over 2000 years from Hippocrates to the 19th century for vague and incorrect notions to be replaced with objective testable scientific concepts which form the basis of modern management.
2. Epidural hematomas (EDHs) require an initial dura separation to form.
3. Secondary volume increase from ongoing arterial bleeding occurs.
   a. The area of initial dura separation required is not great.
   b. The initial limited dura separation is enough to permit ongoing bleeding with a further hematoma volume increase.
   c. The volume of observed clinical hematomas requires secondary dura separation.
   d. Much of the dura is attached loosely over the area where EDHs accumulate.
4. Secondary expansion of a hematoma cannot be due to bleeding with a venous origin.
   a. Bleeding at surgery is from an open skull and says little about what is possible with a closed skull where the resistance to bleeding is too great for a venous bleeding source to separate the dura.
   b. Probably a venous source can fill an initial dura separation pocket but this is no direct evidence.
5. The dural and cranial veins provide a shunt through which epidural blood can leave the skull.
   a. This mechanism provides a mechanism whereby bleeding may be prolonged in the presence of a persisting conscious patient.
6. A small EDH requires no treatment.

Intracranial Epidural Bleeding. DOI: https://doi.org/10.1016/B978-0-12-812159-7.00017-5

7. A lethal EDH causes tentorial herniation with secondary consequences.
    a. There is progressive increasing supratentorial pressure with reduced perfusion pressure.
    b. A potentially lethal cerebral ischemia is found above the tent and into the upper pons.
    c. There is no ischemia in the medulla oblongata and no foramen magnum herniation.
    d. Secondary hydrocephalus due to aqueduct obstruction due to tentorial herniation increases the supratentorial volume load together with ischemic induced cytotoxic edema.
8. Lethal EDH is concluded with a Cushing response started with bradypnea, and hypoxia which induces hypertension and bradycardia followed finally with cardiovascular collapse.
9. Treatment includes observation following computed tomography (CT) in suspected cases.
    a. Serious life-threatening EDHs are treated by surgery and hemostasis.
        i. Their acute management should involve intubation, adequate ventilation, and maintenance of an adequate blood pressure with avoidance of hypotension.
    b. Smaller hematomas may be observed.

QUO VADIS

The author well remembers during an interview for a post as a senior resident being asked what he thought about the future of neurosurgery. It seemed to him at the time an unnecessary temptation to uninformed guess work though that was not something that could be said to senior colleagues in the tense milieu of an interview.

So where are we going with the management of EDH. The earlier list indicates that basically the treatment is well understood and few improvements are to be expected in that area. The obvious difficulty as indicated in Chapter 9, The 20th Century, is making the decision on which patients should be observed and which sent home. None of the clinical indicators in this respect are entirely satisfactory.

The difficulty is that EDH is treatable and if left untreated finally deadly often with a rapid preterminal course. In consequence, those considered to be at risk are observed overnight which is costly and not always easy to implement efficiently. The alternative of performing a CT on every patient is theoretically attractive but wholly impractical. There would be far too many patients for existing facilities and the cost of the examination in which only a tiny minority would show a treatable hematoma would be prodigiously expensive, particularly set against the limited benefit.

The social and geographical specifics of different locations preclude a one solution fits everyone arrangement. The author suggests that every effort should be made to make existing arrangements of observation and transport work as well as possible. However, notable improvements in management are unlikely until miniaturization and future cost reduction makes the development of a portable imaging unit universally available. This would surely solve the problems inherent in EDH treatment. At the present the development of such technology seems out of reach. However, if the last 50 years have taught us nothing else, we should have learned that we are constantly underestimating the speed and power of technological development. That thought may not give us an answer just yet but it can be a basis for cautious optimism.

INDEX

Note: Page numbers followed by "*f*" and "*t*" refer to figures and tables.

–

Printed in the United States
By Bookmasters